Getting the Facts Straight

Getting the Facts Straight

A Viewer's Guide to PBS's *Evolution*

Seattle   Discovery Institute Press   2001

This *Viewer's Guide* is a response to the television documentary *Evolution* that first aired on the PBS network September 24–27, 2001. That series was co-produced by the WGBH/NOVA Science Unit and Clear Blue Sky Productions. This response was developed by the Discovery Institute of Seattle, Washington and is based on videotapes distributed in advance by the producers of the series.

This book evaluates each of the seven programs in that series from a historical and scientific perspective. It points out areas where the series presents inaccurate history or flawed reasoning. It illustrates how the series takes issues that are vigorously debated within the scientific community and presents them as established facts. It also comments on the religious stereotyping engaged in by the producers.

Library Cataloging Data
Getting the Facts Straight: A Viewer's Guide to PBS's "Evolution"
155 p. 23 cm.
Includes 2 illustrations, 8 teaching activities, and endnotes
ISBN: 0-9638654-5-5
LC: 2001095464
Keywords: Evolution; Intelligent Design; Science; Religion; Darwinism
Published in the United States of America on acid-free paper.
First Edition: September 2001
Internet: http://www.ReviewEvolution.com and www.discovery.org

Table of Contents

Episodes

Dramatizations of Darwin's life, interspersed with commentaries by philosopher Daniel Dennett, biologist Stephen Jay Gould, and historian James Moore. Darwin's theory that all living things evolved from a common ancestor in one "great tree of life." Drug resistance in HIV. Biologist Kenneth Miller on the vertebrate eye and the role of God. Similarities between humans and apes.

Fossil whales. Similarities in limb bones. The transition from water to land animals. A "genetic toolkit" common to all animals. The transition from apes to humans.

Mass extinctions. Dinosaurs and the first mammals. Saving an unspoiled forest near Bangkok. Biological invaders in the Hawaiian Islands. Using a beneficial insect to control weeds in North Dakota.

Multi-drug resistant tuberculosis in Russia. Lessons from a cholera epidemic. Feline immunodeficiency virus. Leaf-cutter ants and symbiosis. Allergies and the importance of interactions among species.

Genetic variability as a reason for sexual reproduction. Sexual selection and peacocks' tails. Lessons from chimpanzees and bonobos. Evolutionary psychology on the role of sex in brain evolution.

The emergence of art, technology, and society about 50,000 years ago. Hominid evolution and Neanderthals. Early human migrations. Language. Memes and how they now counteract biological evolution.

The creation-evolution controversy and U.S. science education. Biblical literalist Ken Ham. Students at Wheaton College struggle with their faith. A school board denies a petition to teach special creation alongside evolution.

Activities

Discovery Institute

Discovery Institute is a non-profit, non-partisan public policy center for national and international affairs. The Institute's mission is to make a positive vision of the future practical. The Institute discovers and promotes ideas in the common sense tradition of representative government, the free market and individual liberty. Its mission is promoted through books, reports, legislative testimony, articles, public conferences and debates, plus media coverage and the Institute's own publications and website.

Current projects explore the fields of technology, science and culture, reform of the law, national defense, the environment and the economy, the future of democratic institutions, transportation, religion and public life, government entitlement spending, foreign affairs and cooperation within the bi-national region of "Cascadia."

Contact Information

Discovery Institute
1402 Third Ave Suite 400
Seattle, WA 98101
(206) 292-0401
fax (206) 682-5320
http://www.discovery.org

PBS's *Evolution* Spikes Contrary Scientific Evidence, Promotes Its Own Brand of Religion

This executive summary introduces the more extensive *Viewer's Guide*. It is intended for those who are unable to read the entire guide, but it should not be considered a complete summary.

Accuracy and objectivity are what we should be able to expect in a television documentary—especially in a science documentary on a publicly-funded network. But the PBS series *Evolution* distorts the scientific evidence and promotes a biased religious agenda, thereby betraying our expectations and violating PBS's own official policies.

There are many problems with the *Evolution* series. Although some segments are interesting, others just drag, and many are strangely irrelevant to the educational case they purport to be making. The series almost totally ignores the growing body of scientists who contend that Darwinism is in trouble with the evidence, and it repeatedly dismisses all critics of Darwinism as biblical literalists. This is not an objective documentary, but a one-sided piece of advocacy, unworthy of a publicly funded broadcast network.

Major shortcomings of *Evolution* include:

- Its failure to present accurately and fairly the scientific problems with the evidence for Darwinian evolution.

- Its systematic omission of disagreements among evolutionary biologists themselves about central claims in the series, and its complete failure to report the views of scientists who dispute Darwinism at its roots.

- Its excessive and biased focus on religion, despite its insistence to be about science rather than "the religious realm."

- Its inappropriate use by PBS, a government-funded agency, to organize and promote a controversial political action agenda.

A. Scientific Inaccuracy

The failure to present accurately and fairly the scientific problems with the evidence for Darwinian evolution.

"Evolution affects almost every aspect of human life," claim the series producers, "from medicine to agriculture to a person's choice of mate." The seven episodes supposedly present "the underlying evidence" for this contention, yet some of the evidence presented in the series is known to be false, and the remaining evidence provides surprisingly little support for Darwin's theory.

We are told that "powerful evidence" for the common ancestry of all living things is the universality of the genetic code. The genetic code is the way DNA specifies the sequence of proteins in living cells, and *Evolution* tells us that the code is the same in all living things. But the series is badly out of date. Biologists have been finding exceptions to the universality of the genetic code since 1979, and more exceptions are turning up all the time. In its eagerness to present the "underlying evidence" for Darwin's theory, *Evolution* ignores this awkward—and potentially falsifying—fact.

Evolution also claims that all animals inherited the same set of body-forming genes from their common ancestor, and that this "tiny handful of powerful genes" is now known to be the "engine of evolution." The principal evidence we are shown for this is a mutant fruit fly with legs growing out of its head. But the fly is obviously a hopeless cripple—not the forerunner of a new and better race of insects. And embryologists have known for years that the basic form of an animal's body is established before these genes do anything at all. In fact, the similarity of these genes in all types of animals is a problem for Darwinian theory: If flies and humans have the very same set of body-forming genes, why don't flies give birth to humans? The *Evolution* series doesn't breathe a word about this well-known paradox.

Most of the remaining evidence in *Evolution* shows minor changes in existing species—such as the development of antibiotic resistance in bacteria. Antibiotic resistance is indeed an important medical problem, but changes in existing species don't really help Darwin's theory. Such changes had been observed in domestic breeding for centuries before Darwin, but they had never led to new species. Darwin's theory was that the natural counterpart of this process produced not only new species, but also fundamentally new forms of organisms. *Evolution* has lots of interesting stories about scientists studying changes within existing species, but it provides no evidence that such changes lead to new species, much less to new forms of organisms. Nevertheless, it manages to give the false impression that Darwin's theory has been confirmed.

More details on problems with the evidence for evolution presented in this series—including citations to the relevant scientific literature—can be found in the *Viewer's Guide* and its accompanying educational activities.

B. Omission of Disagreement

The systematic omission of disagreements among evolutionary biologists themselves about central claims in the series, and the complete failure to report the views of scientists who dispute Darwinism at its roots.

"For all of us, the future of religion, science and science education are at stake in the creation-evolution debate," the series' narrator declares. But if the "debate" is so important, why is there such an effort to allow only one scientific point of view to be heard in the series itself? *Evolution* starts right off by giving us the false impression that the only opposition to Darwinian evolution in the nineteenth century was religiously motivated. In fact, much of the opposition to it came from scientists. While most scientists became persuaded that some kind of evolution occurred, many of them disputed Darwin's claim that it was driven by an unguided process of natural selection acting on random variations. Instead, leading scientists advocated a type of guided evolution that flatly contradicted Darwin's core thesis. Because of such scientific criticism, according to historian Peter Bowler, Darwin's theory of evolution by natural selection "had slipped in popularity to such an extent that by 1900 its opponents were convinced it would never recover." The makers of *Evolution* have ignored this rich and fascinating history.

Much of the remainder of the series consists—not of evidence—but of interviews with evolutionary theorists giving us their interpretations of a few ambiguous facts. And surprisingly, the series completely ignores biologists who—though strongly committed to Darwinian evolution—are also strongly critical of the interpretations being presented.

For example, several episodes deal with human origins. We are treated to lots of wildlife photography of apes, and numerous dramatizations featuring human actors in "missing link" costumes, seen from afar—like shots of "Bigfoot"—while we listen to stories told by people who apparently think a very little evidence can go a very long way. But Henry Gee, chief science writer for *Nature* (and an evolutionist), has pointed out that all the evidence for human evolution between about 10 and 5 million years ago "can be fitted into a small box." According to Gee, the conventional picture of human evolution as lines of ancestry and descent is "a completely human invention created after the fact, shaped to accord with human prejudices." Putting it even more bluntly, Gee wrote in 1999: "To take a line of fossils and claim that they represent a lineage is not a scientific hypothesis that can be tested, but an assertion that carries the same validity as a bedtime story—amusing, perhaps even instructive, but not scientific."

The makers of *Evolution* ignore such in-house critics, preferring to leave viewers with the misleading impression that the evidence for human evolution is much stronger than it really is.

Similar censorship of in-house controversies marks Episodes Five and Six, which deal with the role of sex and the evolution of mind. These episodes rely primarily on interviews with proponents of a controversial new field called "evolutionary psychology." But Jerry Coyne, an evolutionary biologist at the University of Chicago, has written that "evolutionary psychologists routinely confuse theory and speculation"—forget about evidence! Coyne compares evolutionary psychology to now-discredited Freudian psychology: "By judicious manipulation, every possible observation of human behavior could be (and was) fitted into a Freudian framework. The same trick is now being perpetrated by the evolutionary psychologists. They, too, deal with their own dogmas, and not in propositions of science."

So the makers of *Evolution* have effectively censored important controversies within the field of evolutionary biology. They have thereby missed a golden opportunity to make science more interesting for the general public. They have also left viewers with a one-sided and misleading view of what evolutionary biology means to its own practitioners.

But the sin of omission goes much deeper. The series also completely ignores the growing number of scientists who think that Darwinian theory at its root is inconsistent with the latest developments in biochemistry, paleontology, embryology, genetics, information theory, and other fields. According to these scientists, Darwin's unguided process of random variation and natural selection is insufficient to account for the highly ordered complexity found in biological systems, which show evidence of directed development or "intelligent design." (Contrary to the Darwinist claim, intelligent design theorists do not claim that science can show us the identity of a designer.)

Scientists advocating a design approach include professors at a number of colleges and universities. The producers of *Evolution* are very aware of this large and growing movement. This is clear from the background materials distributed to PBS affiliates, which include answers to anticipated challenges from intelligent design scholars.

Early efforts to persuade the producers to include scientific critics of Darwinism in the body of the series were rebuffed. Instead, the producers invited some of these critics to come on camera to tell their "personal faith stories" for the last program (Episode Seven), "What About God?" In this way, all critics of Darwinian evolution could be portrayed as religiously motivated. Scientists who criticize Darwinism from an intelligent design perspective did not want to contribute to this misleading stereotype, and so refused to be interviewed for this episode.

By suppressing real disagreements among evolutionary biologists, and by ignoring scientists who think that Darwin's theory is fundamentally flawed, the makers of *Evolution* present viewers with a picture that is more like propaganda than honest journalism. Instead of reporting about evolution, which would include coverage of the theory's problems and critics, the producers of the series present a one-sided advocacy of Darwinism, treating their Darwinian brand of evolutionary theory like an infallible religious dogma. Indeed, they refuse to grant that even a single fact exists that might not corroborate Darwin's theory, insisting that "all known scientific evidence supports evolution." This dogmatic attitude is completely at odds with the spirit of scientific inquiry. By treating Darwin's theory as something that is beyond criticism or contrary evidence, *Evolution* leaves viewers with a shallow and misleading understanding of how science is supposed to work.

C. Religious Bias

The excessive and biased focus on religion, despite the series' insistence to be about science rather than "the religious realm."

According to the producers, "the *Evolution* project presents facts and the accumulated results of scientific inquiry; which means understanding the underlying evidence behind claims of fact and proposed theories. . . . In keeping with solid science journalism we examine empirically-testable explanations for 'what happened,' but don't speak to the ultimate cause of 'who done it'—the religious realm."

Yet the series speaks to the religious realm from start to finish. Episode One is organized around a fictionalized account of Darwin's life, which begins with a scene pitting Charles Darwin, the enlightened scientist, against Captain Robert FitzRoy, the supposed religious fundamentalist. In fact, however, the two men shared similar views when Darwin sailed with FitzRoy aboard the HMS *Beagle*, because Darwin at that time in his life was more religious and FitzRoy was more scientific than this scene implies. Distorting the historical facts, this scene serves to set the stage for all that follows by casting everything in the stereotype of scientist versus religious fundamentalist.

This first episode takes its name, "Darwin's Dangerous Idea," from a book by philosopher Daniel Dennett. Dennett regards Darwinism as a "universal acid" that eats through virtually all traditional beliefs—especially Christianity—and he tells us that Darwin's theory of evolution by natural selection was "the single best idea anybody ever had." People "used to think of meaning coming from on high and being ordained from the top down," Dennett says, but we must now "replace the traditional idea of God the creator with the idea of the process of natural selection doing the creating."

We subsequently meet biologist Kenneth Miller sitting in a church, where he says: "I'm an orthodox Catholic and I'm an orthodox Darwinist." He later explains that "if God is working today in concert with the laws of nature, with physical laws and so forth, He probably worked in concert with them in the past. In a sense, in a sense, He's the guy who made up the rules of the game, and He manages to act within those rules." Yet we are given no hint of the great range of religious views between that of the Bible-thumping FitzRoy and the evolution-friendly Miller. The episode concludes with historian James Moore, who tells us that "Darwin's vision of nature was, I believe, fundamentally a religious vision."

Subsequent episodes include religious imagery such as Michelangelo's Sistine Chapel painting of God touching Adam (while the narrator informs us that our origin, despite the painting, was really not special), and religious music such as the *kyrie eleison* from an African mass (while we watch actors presumably playing our ancestors walk across an African plain). And if we had any doubts that the message of *Evolution* is fundamentally about religion, those doubts are dispelled in the final episode, "What About God?"

"The majesty of our Earth, the beauty of life," Episode Seven begins, "are they the result of a natural process called evolution, or the work of a divine creator?" We are taught that ignorant biblical literalists are the only people who reject Darwinian evolution, and that people who want to sneak religion into the science classroom often intimidate or censor Darwinists. Nothing is said about the many critics of Darwinian evolution who are not even Christians, much less biblical literalists. And nothing is said about the growing number of cases in America in which advocates of Darwinism intimidate and censor their scientific critics.

We are also told in this episode that U.S. science education was "neglected" between the 1925 Scopes trial and the 1957 launch of Sputnik, because evolution was "locked out of America's public schools" during those decades. We are supposed to believe that religious opponents of Darwinism stunted scientific progress. Yet American schools during those supposedly benighted decades produced twice as many Nobel Prize-winners in physiology and medicine as all other countries in the world put together.

Although the producers of *Evolution* promised not to speak to the religious realm, they speak to it forcefully and repeatedly. The take-home lesson of the series is unmistakably clear: Religion that fully accepts Darwinian evolution is good. Religion that doesn't is bad. Now, the producers of *Evolution* are entitled to their opinion. In America, everyone is. But why is this opinion presented as science, on publicly supported television?

D. Promoting a Political Agenda

The inappropriate use of the series by PBS, a government-funded agency, to promote a controversial political agenda.

PBS is funded in part by American taxpayers. As a government-funded agency, it is supposed to be held to high standards of fairness. It is absolutely inappropriate for PBS to engage in activities designed to influence the political process by promoting one viewpoint at the expense of others. Yet an internal document prepared by the Evolution Project/WGBH Boston shows that those behind the series are trying to do just that. Sent to PBS affiliates during the summer of

2001, the document outlines the overall goals for the PBS series and describes its marketing strategy.

According to the document, one of the goals is to "co-opt existing local dialogue about teaching evolution in schools." Another goal is to "promote participation," including "getting involved with local school boards." Moreover, "government officials" are identified as one of the target audiences for the series, and the publicity campaign accompanying the series will include the writing of op-eds and "guerilla/viral marketing." Clearly, one purpose of *Evolution* is to influence Congress and school boards and to promote political action regarding how evolution is taught in public schools.

The political agenda behind *Evolution* is made even more explicit by its enlistment of Eugenie Scott as one of the official spokespersons for the project. Scott runs the National Center for Science Education (NCSE), an advocacy group that by its own description is dedicated to "defending the teaching of evolution in the public schools." As a crucial propaganda tool, the NCSE routinely lumps together all critics of Darwinism as "creationists." According to the group's web site, the NCSE provides "expert testimony for school board hearings," supplies citizens with "advice on how to organize" when "faced with local creationist challenges," and assists legal organizations that litigate "evolution/creation cases." It is a single-issue group that takes only one side in the political debate over evolution in public education. It is therefore completely inappropriate for PBS to enlist NCSE's executive director as an official spokesperson on this project—while excluding other views.

Imagine, for a moment, that PBS created a seven-part series on abortion that was designed to "co-opt existing local dialogue" about abortion legislation, and to influence national and local government officials regarding abortion legislation. Imagine further that PBS defended only one viewpoint in the abortion debate, and enlisted as an official spokesperson the head of a major lobbying group promoting that viewpoint. Would anyone think this was either appropriate or fair?

E. Summary

In summary, the PBS *Evolution* series distorts the scientific evidence, omits scientific objections to Darwin's theory, mischaracterizes scientific critics of Darwinism, promotes a biased view of religion, and takes a partisan position in a controversial political debate. By doing this, PBS has forsaken objectivity, violated journalistic ethics, and betrayed the public trust. It is for these reasons that we have prepared *Getting the Facts Straight: A Viewer's Guide to PBS's "Evolution."*

Note: Quotations from the producers about their goals are taken from "The Evolution Controversy: Use It Or Lose It"—a document prepared by Evolution

Project/WGBH Boston and distributed to PBS affiliates on June 15, 2001. The document concludes by suggesting that "any further questions" should be directed to WGBH at http://www.wgbh.org/. The quotation from Peter Bowler is from his book *Evolution: The History of an Idea* (Berkeley: University of California Press, 1989). Quotations from Henry Gee are from his book, *In Search of Deep Time* (New York: The Free Press, 1999). Quotations from Jerry Coyne are from his book review, "Of Vice and Men," from *The New Republic* (April 3, 2000).

INTRODUCTION

Getting the Facts Straight: A Viewer's Guide to PBS's *Evolution*

Our Web Site: www.ReviewEvolution.com/

The controversy over Darwin's theory of evolution has never been more intense. The American people—and especially America's students—deserve to know what the fuss is all about. They deserve to know what the evidence shows, what scientists really think, and why—after all these years—there is still widespread opposition to Darwinian evolution.

American public television can and should be used to educate people about this important controversy. The seven-part *Evolution* series, produced for public television by Clear Blue Sky Productions and the WGBH/NOVA Science Unit, could have been an important contribution in this regard. But *Evolution* is a work of advocacy, an advertisement not just for Darwinism, but for some of its more extreme manifestations. It distorts the biological evidence, mischaracterizes historical facts, ignores series disagreements among evolutionary biologists themselves, and misrepresents Darwin's scientific critics in order to convince the American people that evolution is absolutely true and indispensable to our daily lives.

This *Viewer's Guide* has been prepared to correct this one-sided presentation. Where *Evolution* distorts or ignores the facts, this Guide supplies them. Where *Evolution* ignores or misrepresents its critics, this Guide lets them speak for themselves. Although *Evolution* promotes the stereotype that all opponents of Darwin's theory are biblical literalists, this Guide was not written to defend biblical literalism but to defend honest science. It is simply based on the premise that the American people deserve to hear the truth—especially from the television network that they are supporting with their tax money.

A. The Goals of the *Evolution* Series

According to *Evolution*'s producers, their guiding vision has been to convey "the importance of evolution" to a general viewing audience. "Evolution affects almost every aspect of human life," the producers believe. "From medicine to agriculture to a person's choice of mate, evolution touches our daily lives in extraordinary ways. Having a grounding in evolution is key to our understanding of so many issues around us."

The program is billed as straightforward science. "Evolution is a scientific concept, and this is a science series," the producers explain. "The Evolution project presents facts and the accumulated results of scientific inquiry; which means understanding the underlying evidence behind claims of fact and proposed theories, and reporting on those areas where the science is sound. We have enlisted the top minds in all of the sciences to help us present the best scientific understanding of the explanation of life on Earth. In keeping with solid science journalism we examine empirically-testable explanations for 'what happened,' but don't speak to the ultimate cause of 'who done it'—the religious realm."[1]

To summarize, the goals of the *Evolution* series are:

1. To show that evolution is important to "almost every aspect of human life," especially medicine, agriculture, and choice of a mate.

2. To present "the underlying evidence behind claims of facts and proposed theories."

3. To report on "areas where the science is sound."

4. In keeping with "solid science journalism," to examine "empirically-testable explanations" while avoiding "the religious realm."

It will be helpful to keep these in mind while viewing the episodes, and to compare the contents of each episode with the producers' announced goals.

B. How To Use This Guide

The best way to use this *Guide* is to read the chapter about each episode *before* viewing it—though reading the *Guide* after an episode will still be useful. For easier reading the chapters are divided into sections, though the sections do not necessarily correspond to actual segments within the episodes.

The series consists of seven episodes. The first is two hours long, while all the others are one hour long, making a total of eight hours. This *Guide* includes a chapter for each episode, providing a detailed description of its contents—including some verbatim quotations from the narrator or interviewees—and critical comments on how the episode misleads the viewer. Each chapter concludes with notes containing additional information and resources for viewers who want to pursue selected topics in more depth.

For educators who want to use the *Evolution* series as a teaching tool—especially to teach critical thinking skills—an appendix to this *Guide* contains several classroom-ready activities. Because the activities and assignments are all fairly involved, and range in difficulty from fairly simple to advanced, the best way to use them is to choose one or two that seem appropriate for a specific group of students. Alternatively, different students could be assigned (or allowed to choose) different activities—though this would require modifying those activities that involve whole-class participation.

This *Viewer's Guide* was prepared using a pre-release version of the seven-part series shown to journalists during the summer of 2001. It is possible that some of the material presented here—including specific quotations—may differ slightly from the series aired on public television.

Notes

1. Quotations from the producers about their goals are taken from "The Evolution Controversy: Use It Or Lose It"—a document prepared by Evolution Project/WGBH Boston and distributed to PBS affiliates on June 15, 2001. The document concludes by suggesting that "any further questions" should be directed to WGBH. The web site for WGBH is http://www.wgbh.org/

   Related web sites include: http://www.pbs.org/wgbh/nova/

   http://www.pbs.org/

   Other contact information:

   WGBH

   125 Western Avenue

   Boston, MA 02134

   (617) 300-2000

   (617) 300-5400

1

Darwin's Dangerous Idea

Dramatizations of Darwin's life, interspersed with commentaries by philosopher Daniel Dennett, biologist Stephen Jay Gould, and historian James Moore. Darwin's theory that all living things evolved from a common ancestor in one "great tree of life." Drug resistance in HIV. Biologist Kenneth Miller on the vertebrate eye and the role of God. Similarities between humans and apes.

A. The Voyage of the Beagle

Charles Darwin originally tried to follow a family tradition of studying medicine, but he found it not to his liking and switched to divinity school instead. Darwin possessed an abiding interest in nature, however, and in 1831 he took a position as ship's naturalist on board H.M.S. *Beagle,* which the British navy was sending to chart the waters off South America. Darwin also served as a traveling companion for the ship's captain, Robert FitzRoy.

In the PBS series, the curtain rises on a dramatization of Charles Darwin and Captain Robert FitzRoy purchasing a mammal fossil from some herdsmen in South America in 1833. As Darwin and FitzRoy complete the purchase, an exchange between them sets the stage for the entire series:

Darwin: [cleaning a fossil skull] "I wonder why these creatures no longer exist?"

FitzRoy: "Perhaps the ark was too small to allow them entry, and they perished in the flood."

Darwin: [laughs]

FitzRoy: "What is there to laugh at?"

Darwin: "Oh, nothing, nothing."

FitzRoy: [indignant] "Do you mock me, or the Bible?"

Darwin: "Oh, neither."

FitzRoy: [scowling] "What sort of clergyman will you be, Mr. Darwin?"

Darwin: [smiling] "Dreadful. Dreadful."

This dramatization distorts the historical facts. Although FitzRoy was raised in a very religious household, while voyaging on the *Beagle* his views were much closer to Darwin's than this fictitious exchange implies—both because FitzRoy was not so literalistic, and because Darwin was not so skeptical. It was only several years later that FitzRoy took to defending a literal interpretation of Genesis against Darwin's views.

By reading FitzRoy's later views back into this period, the *Evolution* project starts right off by promoting a stereotype that will re-appear throughout the series: Rational scientists accept the evidence in order to understand the reality of the natural world, but they are opposed by irrational fundamentalists who reject the evidence in order to preserve a literal interpretation of the Bible.

The truth is that Darwin's theory was opposed in the nineteenth century by many eminent scientists. While most scientists became persuaded that some kind of evolution occurred, many of them disputed Darwin's claim that it was driven by an unguided process of natural selection acting on random variations. Instead, leading scientists advocated a type of guided evolution that flatly contradicted Darwin's core thesis. Because of such scientific criticism, according to historian Peter Bowler, Darwin's theory of evolution by natural selection "had slipped in popularity to such an extent that by 1900 its opponents were convinced it would never recover." In addition to being opposed by scientists, Darwin's theory was opposed by a broad spectrum of religious believers. But the makers of *Evolution* simply ignore this rich and fascinating history.

In the next scene, Darwin rolls his eyes skeptically in his cabin as FitzRoy reads from the Book of Genesis on the deck above. In actuality, however, Darwin regularly attended the shipboard worship services conducted by FitzRoy. Like the preceding fictionalized scene, this one distorts the historical facts in order to promote the scientist-vs.-fundamentalist stereotype. And both scenes put the lie to *Evolution*'s claim to be only about science, not religion.[1]

B. Darwin's Dangerous Idea

From the *Beagle* in the 1830s the scene shifts to the present, for an interview with Tufts University philosopher Daniel C. Dennett: "If I were to give a prize for the single best idea anybody ever had, I'd give it to Darwin for the idea of natural selection—ahead of Newton, ahead of Einstein. Because his idea unites the two most disparate features of our universe: The world of purposeless, meaningless matter-in-motion, on the one side, and the world of meaning, and purpose, and design on the other. He understood that what he was proposing was a truly revolutionary idea."

This entire two-hour episode is named after Dennett's 1995 book, *Darwin's Dangerous Idea*, which compares Darwin's theory to a "universal acid" that "eats through just about every traditional concept"—including the concept of God. Dennett's book also calls anyone who rejects Darwin's theory "inexcusably ignorant."[2]

The interview with Dennett is followed by one with Harvard University evolutionary biologist Stephen Jay Gould, who says: "The Darwinian revolution is about who we are—it's what we're made of, it's what our life means insofar as science can answer that question. So it, in many ways, was the singularly deepest and most discombobulating of all discoveries that science has ever made."

Following Gould, Open University historian James R. Moore adds: "In Darwin's day the idea of evolution was regarded as highly unorthodox, because it went against all of natural history in Great Britain. It jeopardized the standing of science, it did jeopardize the standing of a stable society, the Bible, and the Church as well." We will meet all three of these men several more times in this episode.

Evolution's producers claim that they "examine empirically-testable explanations for 'what happened,' but don't speak to the ultimate cause of 'who done it'—the religious realm." But Dennett and Gould address issues of meaning and purpose that are normally considered to be in the province of religion, and Moore describes the challenge Darwin posed to society and religion. It seems that *Evolution* has things to say about the religious realm after all.

C. The Legend of "Darwin's Finches"

The interviews with Dennett, Gould and Moore are followed by some more imaginary scenes with Charles Darwin, his brother Erasmus, Captain FitzRoy and others—many serving to reinforce the scientist-vs.-fundamentalist stereotype.

At one point Charles Darwin meets with ornithologist John Gould, who informs him that some birds Darwin had collected on the Galápagos Islands (600 miles west of Ecuador) were finches. Darwin later suggests to FitzRoy that some finches had been blown to the islands from the South American mainland and then diverged into the separate species now present.

In a subsequent scene, John Gould tells Darwin that the Galápagos finches he collected differ mainly in the size and shape of their beaks. Darwin cradles one bird in his hand and remarks: "And they're all descended from this one—the common ground finch!" Darwin's eyes light up and he rushes out of the room to tell his brother that he has finally put the pieces together. "Perhaps everything is part of one ancestral chain," Darwin concludes, and the finches are simply one branch on the great "tree of life."

The scene then shifts to a rainforest in present-day Ecuador, where Boston University biologist Chris Schneider tells us: "One of the most important ideas that Darwin had was that all living things on Earth were related. How can you realize that you are part of this single tree of life and not be fundamentally moved by that? It's something that stirs the soul."

Schneider and his colleagues are studying various animals, the narrator tells us, to "understand how changing environments might trigger the evolution of new species." One of Schneider's colleagues studies differences in birds' beaks, and the narrator explains: "Even subtle differences may offer clues as to how and why new species arise—just as it was the beaks of finches from the nearby Galápagos Islands that spurred Darwin's thinking in the 1830s. Darwin saw that the finches he brought back had uniquely shaped beaks adapted to the different foods on the islands. He envisioned that these different species of finch had all descended with modifications from a common ancestral population that had flown over from the mainland. Darwin's bold insight was to apply this vision to all of life."

This story of "Darwin's finches" is re-told in many biology textbooks, but it is largely fictional. In fact, the Galápagos finches had almost nothing to do with the process by which Darwin arrived at his theory. Much of his information about the birds was erroneous, and since he never visited South America north of Peru he was unaware of differences between the Galápagos birds and those on the mainland. Darwin did not even mention the finches in *The Origin of Species.*

It wasn't until a century after his voyage on the *Beagle* that variations in the birds' beaks were correlated with different food sources, and only then were they named "Darwin's finches" in his honor. Although other Galápagos animals impressed Darwin, the finches did not, and the account presented in this episode is more legendary than historical. This may seem like a minor point, but it is symptomatic of a tendency among Darwin's admirers to give the man credit for things he never did.[3]

D. Darwin's Tree of Life

According to the narrator, Darwin's "bold insight" is "now the bedrock of biology. All forms of life on Earth have evolved from a single branching tree of life." Darwin saw "that the great variety of life on Earth—leopards and lichens, minnows and whales, flowering plants and flatworms, apes and human beings— all descended from one root, one common ancestor."

Enter Stephen Jay Gould again: "It was, indeed, another one of his radical proposals not only to say that evolution happened, but that there was a root, a common ancestry, to everything that lived on this planet—including us. You could construe it in another way, that is (I like to say) more user-friendly: You could have thought, well, God had several independent lineages and they were all moving in certain pre-ordained directions which pleased His sense of how a uniform and harmonious world ought to be put together—and Darwin says, No, it's just history all coming [through] descent with modification from a single common ancestry."

James Moore makes another appearance, declaring: "The key to Darwin's thought in every realm is that given enough time, and innumerable small events, anything can take place by the laws of nature." This includes the raising of mountains and the evolution of new species.

These statements seem strangely out of place here. Why does a program that ostensibly wants to avoid "the religious realm" have Stephen Jay Gould tell us—by his tone, if not in words—that Darwin's theory is preferable to divinely guided separate lineages? And why does a program that has "enlisted the top minds in all of the sciences" rely on a historian to assure us that anything can take place by the laws of nature, given enough time?

The scene now switches back to South America again, where biologists are finding camouflaged insects and measuring the beaks of hummingbirds. The narrator tells us that these birds all evolved from a common ancestor, and that scientists now compare their DNA to determine how long ago they diverged from that ancestor. But in these DNA comparisons common ancestry is simply assumed; where is the *evidence* for it?

And even though the common ancestry of hummingbirds seems plausible, how do we know that "leopards and lichens, minnows and whales, flowering plants and flatworms, apes and human beings" also share a common ancestor? The only actual evidence mentioned in this episode is the supposed universality of the genetic code. According to the narrator: "The fact that the blueprints for all living things are in the same language—the genetic code of DNA—is powerful evidence that they all evolved on a single tree of life."

So the genetic code is supposed to be hard evidence for Darwin's theory that all living things share a common ancestor. Let's take a closer look at it.

DNA is like a string of words that tells a cell how to make the proteins it needs. DNA words, however, are different from the words we use. In English, we use an alphabet of 26 letters, from which we make thousands of words of varying length. DNA, on the other hand, uses an alphabet with only four letters, and it makes only three-letter words—a total of 64 of them. Some of these three-letter words tell the cell to start or stop making a protein, while the others stand for 20 kinds of subunits that the cell uses to assemble it.

DNA words—corresponding to a start or stop signal, or to one of the 20 protein subunits—make up "the genetic code." In the early 1970s, evolutionary biologists thought that a given DNA word specified the same protein subunit in every living thing, and that the genetic code was thus universal. This was unlikely to have happened by chance, so it was interpreted as evidence that every organism had inherited its genetic code from a single common ancestor.

In 1979, however, exceptions to the code were found in mitochondria, the tiny energy factories inside cells. Biologists subsequently found exceptions in bacteria and in the nuclei of algae and single-celled animals. It is now clear that the genetic code is *not* the same in all living things, and that it does *not* provide "powerful evidence" that all living things "evolved on a single tree of life."[4]

So the first hard evidence that we are given for Darwin's tree of life turns out to be false.

The scene in the South American rainforest concludes with Chris Schneider climbing a tower that reaches high into the trees. Schneider asks: "How is it that organisms that are so different can be related—that we are related to a flatworm, or a bacteria? Darwin emphasized that small changes would accrue every generation, and these changes could build up to amount to enormous changes. It's not really hard to understand how major transitions could come about, given that life has been around for three and a half billion years." Schneider concludes by assuring us that "Darwin really had it right."

In science, however, assurances are not enough; we must also have evidence. So far, we have been told that the universality of the genetic code is "powerful evidence" for the relatedness of all living things. But that "evidence" turns out to be false. How about Schneider's assurance that an accumulation of slight differences through natural selection can produce the enormous differences among living things? Where is the evidence for that?

We haven't seen any so far, but let's be patient. Maybe some is on the way.

E. Natural Selection in HIV

The episode moves on to some more dramatizations involving Darwin and his family, and some return appearances by James Moore and Stephen Jay Gould (who outline Darwin's theory of natural selection). We are then treated to some beautiful wildlife photography of animals engaged in the struggle for survival. Despite its faults, the *Evolution* series periodically delights us with colorful and captivating footage of the amazing creatures that inhabit our planet. (The colorful footage, however, is not evidence for evolution.)

The narrator continues: "Darwin couldn't actually see natural selection acting in real time. But today scientists can, by observing the evolution of HIV, the virus that causes AIDS." The scene shifts to an AIDS patient who, we are told, "takes a host of medications, but to little avail. The virus keeps adapting, evolving into new strains that evade the drugs." The scene switches to another AIDS patient, who is also "locked in a daily struggle against the rapidly evolving virus."

According to the narrator, the physician treating the second patient "has seen HIV evolve into new varieties over the last dozen years. The virus is constantly changing, subject to the forces of natural selection, in the environment of the patient's body." This occurs because a drug kills some viruses, but others acquire chance immunity to it and survive. The survivors then resist further treatments by that particular drug.

The attending physician describes this process of evolution: "In the case of HIV, we're talking about minutes to hours to move from one species to another. It's mind-boggling."

It's also **untrue**. The acquisition of drug resistance by HIV is a far cry from moving "from one species to another." The HIV at the end of the treatment is the same species (actually, it's called a "quasispecies") as the HIV at the beginning of the treatment. This is clear from the fact that the virus reverts to its previous condition when treatment is stopped. (Furthermore, HIV reproduces about every 24 hours, so whatever changes are occurring would take days instead of "minutes to hours.")[5]

But changes within a species—such as we observe in the case of HIV—are nothing new. For centuries, farmers have been producing dramatic changes in crops and livestock by artificially selecting certain specimens for breeding. In fact, Darwin and his contemporaries took the success of domestic breeding for granted. But well-bred cows are still cows, and well-bred corn is still corn.

The revolutionary claim in Darwin's theory was that the natural counterpart of artificial selection can create not only new species, but also new forms of organisms. The enormous differences we now see between "leopards and lichens, minnows and whales, flowering plants and flatworms, apes and human beings" go far beyond the small differences we observe in domestic breeding. But artificial selection does not produce new species, much less new forms of organisms, and this has been a major stumbling block to evolutionary theory since the time of Darwin.

The HIV story does nothing to overcome this stumbling block. It does not show the origin of a new species; it shows only the sorts of minor changes *within* species that people have been observing for centuries. So it provides no support for Darwin's theory that natural selection can produce new species and higher level forms.

Enter Stephen Jay Gould and Daniel Dennett again. According to Gould: "All that happens in evolution, at least under Darwinian natural selection, is that organisms are struggling, in some metaphorical and unconscious sense, for reproductive success, however it happens." Dennett adds: "The process of natural selection feeds on randomness, it feeds on accident and contingency, and it gradually improves the fit between whatever organisms there are and the environment in which they're being selected. But there's no predictability about what particular accidents are going to be exploited in this process."

So instead of presenting us with evidence, this episode merely offers us more assurances that natural selection can, indeed, do what Darwin said it could do— and that it's random. The focus shifts back to treating AIDS patients, and how evolutionary thinking allegedly helps them. But the question with which we started was: What is the evidence that natural selection can produce "leopards and lichens, minnows and whales, flowering plants and flatworms, apes and human beings" from one common ancestor? And our question remains unanswered.

F. The Evolution of the Eye

More dramatizations follow, culminating in an important conversation between Darwin and his wife, Emma:

Emma: [reading from the unpublished manuscript of Charles's book, *The Origin of Species*] "You say here that the human eye may possibly have been acquired by gradual selection of slight, but in each case useful, deviations."

Charles: "Yes."

Emma: "That's a very great assumption, Charles."

Charles: "Well, if I'm wrong about that, I'm wrong about everything. My entire theory's in ruins."

Emma: "Can your theory account for the way my eyes, and ears, and hands, and heart combine to reproduce the sounds that Chopin heard in his head? Isn't that a God-given gift?"

Charles: "Well it's given—but not, I think, by God."

James Moore returns and says Emma "saw that her husband's speculations about the origins of species and of humanity would jeopardize the Christian plan of salvation. God was being made remote in her husband's universe. Now if nature, by itself, unaided by God, could make an eye, then what else *couldn't* nature do? Nature could do anything! It could make everything!"

Brown University biologist Kenneth R. Miller elaborates: "In Darwin's day, the very existence of an organ of extreme perfection like the eye was taken by many as proof of God, as proof of a Designer. How else could all the intricate organs and substructures of the eye have come together in just the right way to make vision so possible, and so perfect? But it turns out the eye isn't exactly perfect after all. In fact, the eye contains profound optical imperfections. And those imperfections are proof, in a sense, of the evolutionary ancestry of the eye."

After calling the multi-layered structure of the eye an evolutionary defect (because it can lead to retinal tears as people grow older), the narrator tells us that another defect "occurs because nerve cells and blood vessels evolved to lie in front of the retina, where they interfere with its ability to form sharp images. It's like trying to take a picture through a foggy piece of glass. Now the optic nerve itself evolved to connect to the brain through a hole in the retina. So the eyes of all vertebrates have a small blind spot, right near the middle of the visual field."

Miller attributes this to natural selection: "Evolution starts with what's already there, tinkers with it, and modifies it, but can never do a grand re-design. So even the eye, with all of its optical perfection, has clues to the fact that its origin is of the blind process of natural selection."

But does imperfection count as evidence against design? Many cars are noticeably imperfect, though they are all designed. Every time a manufacturer recalls a faulty product we are confronted with an example of imperfect design. Miller is actually relying here on an unstated **theological** argument (as Darwin often did in *The Origin of Species*)—namely, that if God had made it, it would be perfect; but it's not perfect, so God didn't make it, and evolution must have. But why does an argument for evolution have to resort to theology?

In any case, it turns out that the vertebrate eye is not as imperfect as Miller claims. The light-sensing cells in the eyes of higher vertebrates are extremely efficient at amplifying faint light. The efficient, hard-working tips of the light-sensing cells need lots of energy, and they also need to be constantly regenerated. The energy is provided by a dense bed of capillaries, and the regeneration is facilitated by a special layer of epithelial cells. If the tips of the light-sensing cells faced forward, as Miller thinks they should, incoming light would be blocked by the dense capillary bed and the epithelial layer. Such an eye would be much *less* efficient—and therefore less perfect—than the one we have now, because the capillaries and epithelial cells are now behind the retina instead of in front of it. It's true that the present arrangement causes the optic nerve to leave a blind spot as it passes through the retina; but vertebrates have two eyes, and the blind spots cancel out when both are used to focus on the same object. Despite Miller's claim, the vertebrate eye seems to be a masterpiece of engineering!

Another problem with Miller's argument is its implication that the retina of vertebrate eyes is backwards because evolution was forced to tinker with something it already had. But animals regarded as evolutionarily more primitive than vertebrates all have retinas that face *forward*. There is no backwards retina in a primitive animal that could have served as an evolutionary precursor to the vertebrate eye. So where is Miller's "proof" of the evolutionary ancestry of the vertebrate eye?

Stephen Jay Gould returns to explain that "what Darwin was able to do was to point out that you might think in logic that it's difficult to imagine a set of intermediary stages between the simplest little spot of nerve cells that can perceive light to a lens-forming eye that makes complex images, but in fact these intermediary forms do exist in nature."

We are shown colorful pictures of eyes in various animals, but all of them are complex; none of them are "intermediary forms." The scene then shifts to Sweden, where Lund University zoologist Dan-Eric Nilsson has performed calculations and made a model to show how "we can go all the way gradually, in very small steps, from a simple pigment cup-eye—which has barely got the ability to determine the direction of a light source—all the way to a complete camera-type eye, the same type as we have ourselves."

Nilsson's interesting presentation is periodically interrupted by more pictures of animals, this time showing presumably intermediary forms—including a flatworm with a simple cup eye, a chambered nautilus with a pin-hole camera eye, and some vertebrates. But no biologist believes that chambered nautiluses evolved into vertebrates, so it's not clear what relevance these forms have to the argument. Nevertheless, Nilsson concludes: "And that is really exactly the way eye evolution must proceed."

Nilsson's hypothesis, however, requires a pre-existing layer of light-sensing cells, and these require the simultaneous presence of several extremely complex and specialized molecules. According to Darwin's theory, such a complex molecular apparatus must have formed as a result of innumerable small steps, but no one knows how this could have happened. The origin of the light-sensitive cells that Nilsson needs for his hypothesis remains a mystery.

Furthermore, in order for evolution to work, variations must be heritable—that is, they must be passed on to subsequent generations. There is no known mechanism by which real organisms could generate the variations envisioned by Nilsson, much less pass them on to subsequent generations. So Nilsson's mechanism—plausible though it may seem—has no counterpart in living things.[6]

We are not told about these problems, however, and the narrator concludes: "He [Darwin] later wrote that eyes must have evolved by numerous gradations from an imperfect and simple eye to one perfect and complex, with each grade being useful to its possessor. Nature, unaided by a designer, could produce an organ of seemingly miraculous complexity."

So we are assured that surely eyes *must* have evolved this way. But did they, really?

G. More on Darwinism and Religion

More dramatizations follow, including some poignant scenes about the death of Darwin's daughter, and the effect this tragedy had on his religious faith. Whatever one may think of Darwin's theory, there is no denying that he was an interesting man, and the dramatizations in this episode help to bring him alive for us.

After James Moore and Stephen Jay Gould comment on Darwin's religious views, the scene shifts to a Roman Catholic Church, where Kenneth Miller sits listening to a children's choir sing "All Things Bright and Beautiful." The narrator explains: "Today scientists hold all conceivable views on religion—from atheism, to agnosticism, to a general spirituality. And many, like biologist Ken Miller, adhere to very traditional beliefs."

"I'm an orthodox Catholic, and I'm an orthodox Darwinist," Miller says. "My idea of God is: Supreme being who acts in concert with the principles and the ideas that Darwin explained to us about the origin of species. My students often ask me, You say you believe in God—well what *kind* of God? Is it a fashionable New Age God, a pyramid-power kind of God? Do you think, like some scientists do, that God is the sum total of the laws of physics? And I shake those off and say that my religious belief is entirely conventional." The congregation recites the "Our Father" in the background as Miller continues: "It surprises students very often that anyone could say that, that kind of very traditional, conventional religious belief could be compatible with evolution, but it is. I find this absolutely wonderful consistency with what I understand about the universe from science and what I understand about the universe from faith."

In the next scene, Miller is being interviewed on a radio program in Knoxville, Tennessee, about his recent book, *Finding Darwin's God.* "What room is there for God in present-day life?" Miller asks himself, then answers: "Well, I think if you ask people who are believers, How does God act?—they would say He acts in a variety of ways. He answers our prayers; He inspires us. No doubt there are events that take place that are part of what some people might call God's plan. And what I would suggest is—if you look back in Earth's history—if God is working today in concert with the laws of nature, with physical laws and so forth, He probably worked in concert with them in the past. In a sense, in a sense, He's the guy who made up the rules of the game, and He manages to act within those rules."[7]

The narrator explains: "For Miller, and millions of followers of all major religions, notions of God and evolution are fully compatible. But not everyone agrees." Daniel Dennett comes on again and says: "When we replace the traditional idea of God the creator with the idea of the process of natural selection doing the creating, the creation is as wonderful as it ever was. All that great design work had to be done. It just wasn't done by an individual, it was done by this huge process, distributed over billions of years."

A clergyman then reads to his congregation from the Bible: "God created man in His image, in the image of God He created him; male and female He created them." As the clergyman continues speaking in the background, Dennett remarks: "Whereas people used to think of meaning coming from on high and being ordained from the top down, now we have Darwin saying, No, all of this design can happen, all of this purpose can emerge from the bottom up, without any direction at all. And that's a very unsettling thought for many people."

Some comments by James Moore are followed by more dramatizations featuring Darwin and his contemporaries. These include Darwin's receipt of Alfred Russel Wallace's manuscript outlining a theory very similar to his own (which prompted Darwin finally to publish his theory in *The Origin of Species*), and a re-enactment of the famous 1860 confrontation between Bishop Samuel Wilberforce and Darwin's defender, Thomas Henry Huxley. Though much of the history in these dramatizations is accurate, it downplays the scientific opposition to Darwin's theory and over-emphasizes the religious opposition.

If we had any doubt before, it is now absolutely clear—despite what the producers claim—that the *Evolution* project has quite a lot to say about "the religious realm." After presenting us with a variety of conflicting religious viewpoints, it leaves us with the distinct impression that some of them are acceptable while others are not. How can we tell the difference between the good and the bad? By their attitudes toward Darwinian evolution. Religion that accepts Darwin's theory is good, while religion that doesn't is bad.

So the PBS *Evolution* series is far from neutral on religious issues. It has a very specific agenda when it comes to "the religious realm." That agenda was subtly introduced in the very first scene; it becomes more explicit here; and—as we shall see in the last episode—it turns out to be one of the major take-home lessons of the series.

H. Are Humans Just Animals?

The concluding scenes of this episode tackle the question of whether there is more to human beings than their animal nature. Daniel Dennett addresses the question first: "For more than a century, people have often thought that the conclusion to draw from Darwin's vision is that *Homo sapiens*—our species—that we're just animals, too, we're just mammals, that there is nothing morally special about us. I, in myself, don't think this follows at all from Darwin's vision, but it is certainly the received view in many quarters."

Watch closely now as the camera pans over a stack of books critical of Darwinism. The narrator says: "Ever since *The Origin of Species* was published, strict believers in biblical creation have attacked Darwin's vision. Their concerns aren't only about the science of evolution. At stake, many believe, is nothing less than the human soul." But two of the books in the stack are by critics of evolution who are *not* strict believers in biblical creation. *Darwin On Trial* is by Phillip E. Johnson, a Berkeley law professor who is a Christian but who accepts the geological evidence for an old Earth; and *Darwinism: The Refutation of a Myth* is by Søren Løvtrup, an evolutionary biologist. Yet *Evolution* misrepresents them as biblical literalists simply because they are critical of Darwin's theory.

James Moore continues: "To suggest that animals and plants—and us!—came into being in a natural, law-like way, in the way the planets move, was to put in jeopardy the human soul. And the human soul is the crux of the matter, because if we are not different from animals—if we don't live forever, in heaven, or in hell—then why should we behave other than like animals in this life?"

The scene shifts to chimpanzees. We are told that in Darwin's time there was not much evidence that chimps and humans are closely related, but the fossil record and DNA studies have since shown that they are. Similarity, especially in the DNA, supposedly shows that we come from a common ancestor.

There are similarities in the way we learn, too. Ohio State University developmental psychologist Sally Boysen concludes from this that humans and chimps "have a great deal of commonality in—literally—the neurological structure that supports their ability to learn just like we do. Those things are absolutely comparable, and **had** to come from a common ancestor."

Kenneth Miller returns to the screen to tell us that for all of the extraordinary similarities between us and the apes, "there are striking differences," which he attributes to natural selection. According to Miller, "Darwin's great idea is a grand and marvelous explanation that shows us that we are united with every other form of life on this planet. And I find that an exciting, and maybe even ennobling way, to look at things."

So, are we just animals, built to behave like them—or are we morally special, with immortal souls? Dennett thinks the former doesn't follow from Darwin's theory; Miller, as an orthodox Darwinist and an orthodox Catholic, presumably agrees; and Moore acknowledges that this question is "the crux of the matter." Boysen thinks our similarities with chimps mean we had to come from a common ancestor, but she doesn't comment on the larger question, which remains unanswered.[8]

As we saw above, however, in the scene with the stack of books, anyone critical of Darwinian evolution risks being stereotyped as a strict believer in biblical creation. The message seems to be that it's OK for people to believe whatever they want about God and the soul—as long as they don't criticize evolution. Once again, there's good religion and there's bad religion—and Darwinism enables us to distinguish between the two.

The episode ends, appropriately, with Darwin's death. Moore concludes by telling us that Darwin was accorded great honor because "he had naturalized creation, and had delivered human nature, and human destiny," into the hands of those who were running the technological society of late nineteenth-century England. "Society would never be the same. Darwin's vision of nature was, I believe, fundamentally a religious vision."

Amen.

Notes

1. According to historian Janet Browne, FitzRoy at the time of the *Beagle* voyage was "much inclined to believe [Geologist Charles] Lyell's revisionist anticlerical arguments" *against* the historicity of Noah's Flood:

> In particular, FitzRoy doubted the existence of a Noachian Flood—the traditional stumbling-block for Protestants since the seventeenth century. . . . He could not believe that the extensive gravel and clay deposits of Patagonia, hundreds of feet thick and apparently originating in calm water, had been laid down in forty days. Lyell's secular proposals seemed altogether more probable.
>
> He said as much to Darwin while they tramped the coastal plains puzzling over the sequence of deposition. To understand these deposits was important in relation to the fossil mammals and FitzRoy was as keen to unravel the conundrum as Darwin.
>
> (Janet Browne, *Charles Darwin, Voyaging: Volume I of a Biography* [New York: Knopf, 1995], 272-273)

During the *Beagle* voyage, Darwin's religious views were largely indistinguishable from those of FitzRoy. (Such disagreements as they had concerned the topic of slavery [see Browne, 199] and social niceties.) Indeed, during the voyage, FitzRoy and Darwin composed a joint letter defending the work of English missionaries in Tahiti and New Zealand (Browne, 330). Furthermore, as Browne reports:

> [Darwin] went to church regularly throughout the voyage, attending the shipboard ceremonies conducted by FitzRoy and services on shore whenever possible. He and [naval lieutenant Robert] Hammond, spent some hours in Buenos Aires waiting to hear if they could receive communion from the English chaplain stationed there before going to Tierra Del Fuego. . . . The *Beagle* Darwin, though occasionally doubtful, was by no means a thorn in the side of the church. (326)

According to Gertrude Himmelfarb: "Most biographers, carried away by the zeal of hindsight, tend to hasten the development of Darwin's religious views [i.e., away from orthodoxy], as they do his evolutionary views. . . . [But there is a] total lack of evidence on this score." Furthermore:

> Several times during the voyage he [i.e., Darwin] alluded to the vision of a quiet English parsonage glimpsed through a grove of tropical palms. To a college friend already installed in a country parish he wrote [in Nov. 1832]: "I hope my wanderings will not unfit me for a quiet life and that on some future day I may be fortunate enough to be qualified to become like you a country clergyman. And then we will work together at Natural History." And again the following year [May 23, 1833]: "I often conjecture what will become of me; my wishes certainly would make me a country clergyman."

> (Gertrude Himmelfarb, *Darwin and the Darwinian Revolution* [New York: W. W. Norton, 1959], 64-65, 459.)

The quotation from Bowler is from Peter J. Bowler, *Evolution: The History of an Idea,* Revised Edition (Berkeley: University of California Press, 1989), 246. The scientific opposition to Darwin's theory is described in David Hull's *Darwin and His Critics: The Reception of Darwin's Theory of Evolution by the Scientific Community* (Chicago: University of Chicago Press, 1983), while the broad spectrum of nineteenth-century religious opposition to Darwin is described in James R. Moore's *The Post-Darwinian Controversies* (Cambridge: Cambridge University Press, 1979).

There is no historical evidence for Darwin's shipboard reaction to FitzRoy's Bible reading. According to Himmelfarb (64), Darwin was "shocked when one of his shipmates flatly denied the fact of the flood. Indeed, he was remembered by them as being naively orthodox in his beliefs. Several of the officers (though themselves orthodox) were amused once when he unhesitatingly gave the Bible as final authority on a debated point of morality." Further details of Darwin's life can be found in Adrian Desmond and James Moore, *Darwin: The Life of a Tormented Evolutionist* (New York: W.W. Norton, 1991).

2. Daniel Dennett believes that Darwinian evolution is not only unquestionably true, but also bears "an unmistakable likeness to universal acid—it eats through just about every traditional concept." (*Darwin's Dangerous Idea: Evolution and the Meanings of Life* [New York: Simon & Schuster, 1995], 63) In other words, Dennett sees a conflict

between Darwin's theory and all traditional forms of religion—not just biblical fundamentalism.

Some other quotes from Dennett's book (after which this episode is named) are:

Darwin's dangerous idea cuts much deeper into the fabric of our most fundamental beliefs than many of its sophisticated apologists have yet admitted, even to themselves. (18)

To put it bluntly but fairly, anyone today who doubts that the variety of life on this planet was produced by a process of evolution is simply ignorant—inexcusably ignorant. (46)

Evolutionists who see no conflict between evolution and their religious beliefs have been careful not to look as closely as we have been looking, or else hold a religious view that gives God what we might call a merely ceremonial role to play. (310)

Those whose visions dictate that they cannot peacefully coexist with the rest of us we will have to quarantine as best we can. . . . If you insist on teaching your children falsehoods—that the Earth is flat, that 'Man' is not a product of evolution by natural selection—then you must expect, at the very least, that those of us who have freedom of speech will feel free to describe your teachings as the spreading of falsehoods, and will attempt to demonstrate this to your children at the earliest opportunity. Our future well-being—the well-being of all of us on this planet—depends on the education of our descendants. What, then, of all the glories of our religious traditions? They should certainly be preserved, as should the languages, the art, the costumes, the rituals, the monuments. (519)

Dennett recommends that religion be "preserved in cultural zoos." His book concludes with: "Is something sacred? Yes, I say with Nietzsche. I could not pray to it, but I can stand in affirmation of its magnificence. This world is sacred." (520)

3. According to historian of science Frank Sulloway, Darwin "possessed only a limited and largely erroneous conception of both the feeding habits and the geographical distribution of these birds." And as for the claim that the Galápagos finches impressed Darwin as evidence of evolution, Sulloway wrote, "nothing could be further from the truth." (*Journal of the History of Biology* 15 [1982], 1-53; *Biological Journal of the Linnean Society* 21 [1984], 29-59.)

Darwin wrote in the second edition of his *Journal of Researches* (London: John Murray, 1845, 380): "The most curious fact is the perfect gradation in the size of the beaks of the different species of [finches]. Seeing this gradation and diversity of structure in one small, intimately related group of birds, one might really fancy that from an original paucity of birds in this archipelago, one species had been taken and modified for different ends." But Darwin by then had already formulated his theory, so this was a speculative afterthought. Indeed, the confusion surrounding the geographical labeling of Darwin's specimens (alluded to in the PBS episode) made it impossible for him to use them as evidence for his theory. Nor did Darwin have a clear idea of the finches' ancestry. He did not visit the western coast of South America north of Lima, Peru, so for all Darwin knew the finches were identical to species still living on the mainland.

It wasn't until the 1930s that the Galápagos finches were elevated to their current prominence. Although they were first called "Darwin's finches" by Percy Lowe in 1936 (*Ibis* 6: 310-321), it was ornithologist David Lack who popularized the name a decade later. Lack's 1947 book, *Darwin's Finches* (Cambridge University Press), summarized the evidence correlating variations in finch beaks with different food sources, and argued that the beaks were adaptations caused by natural selection. In other words, it was Lack more than Darwin who imputed evolutionary significance to the Galápagos finches. Ironically, it was also Lack who did more than anyone else to popularize the myth that the finches had been instrumental in shaping Darwin's thinking.

4. The universality of the genetic code was suggested by Francis Crick in "The Origin of the Genetic Code," *Journal of Molecular Biology* 38 (1968), 367-379. Exceptions to the code are reviewed in Syozo Osawa, *Evolution of the Genetic Code* (Oxford: Oxford University Press, 1995), and in Robin D. Knight, Stephen J. Freeland and Laura F. Landweber, "Rewiring the Keyboard: Evolvability of the Genetic Code," *Nature Reviews: Genetics* 2 (2001), 49-58. For a current list of exceptions to the genetic code, go to:

    http://www.ncbi.nlm.nih.gov/htbin-post/Taxonomy/wprintgc?mode=c

5. Swarms of HIV variants—which may include drug-resistant strains—are called "quasispecies." See Esteban Domingo et al., "Basic concepts in RNA virus evolution," *FASEB Journal* 10 (1996), 859-864; and Michael H. Malim and Michael Emerman, "HIV-1 Sequence Variation: Drift, Shift, and Attenuation," *Cell* 104 (2001), 469-471.

6. The idea that the vertebrate eye is imperfect, and therefore must be a product of Darwinian evolution, did not originate with Kenneth Miller. In his 1986 book, *The Blind Watchmaker* (New York: W. W. Norton), Richard Dawkins wrote:

    Any engineer would naturally assume that the photocells would point towards the light, with their wires leading backwards towards the brain. He would laugh at any suggestion that the photocells might point away from the light, with their wires departing on the side nearest the light. Yet this is exactly what happens in all vertebrate retinas. Each photocell is, in effect, wired in backwards, with its wire sticking out on the side nearest to the light. The wire has to travel over the surface of the retina, to a point where it dives through a hole in the retina (the so-called 'blind spot') to join the optic nerve. This means that the light, instead of being granted an unrestricted passage to the photocells, has to pass through a forest of connecting wires, presumably suffering at least some attenuation and distortion (actually probably not much but, still, it is the principle of the thing that would offend any tidy-minded engineer!). (p. 93)

    See also Timothy H. Goldsmith, "Optimization, Constraint, and History in the Evolution of Eyes," *The Quarterly Review of Biology* 65:3 (1990), 281-322.

    For a discussion of the role played in Darwinian thinking by hidden theological arguments (like the argument that God would only make perfect things), see Paul A. Nelson, "The Role of Theology in Current Evolutionary Reasoning," *Biology and Philosophy* 11 (1996), 493-517; and Cornelius George Hunter, *Darwin's God* (Grand Rapids, MI: Brazos Press, 2001).

For arguments against the modern Darwinian claim that the vertebrate eye is imperfect, see George Ayoub, "On the Design of the Vertebrate Retina," *Origins and Design* 17:1 (1997), available at:

http://www.arn.org/docs/odesign/od171/retina171.htm

See also Michael J. Denton, "The Inverted Retina: Maladaptation or Pre-adaptation?" *Origins and Design* 19:2 (1997), available at:

http://www.arn.org/docs/odesign/od192/invertedretina192.htm

For general background on the evolution of eyes, see L. Salvini-Plawen and Ernst Mayr, "On the Evolution of Photoreceptors and Eyes," *Evolutionary Biology* 10 (1977), 207-263. Neuroanatomist Bernd Fritzsch, though an evolutionist, criticizes over-simplified explanations for the evolution of the vertebrate eye in "Ontogenetic Clues to the Phylogeny of the Visual System," in *The Changing Visual System,* edited by P. Bagnoli and W. Hodos (New York: Plenum Press, 1991), 33- 49.

For a published report of Nilsson's work, see Dan-Eric Nilsson and Susanne Pelger, "A pessimistic estimate of the time required for an eye to evolve," *Proceedings of the Royal Society of London* B 256 (1994), 53-58. See also Richard Dawkins, "The eye in a twinkling," *Nature* 368 (1994), 690-691.

The lack of a Darwinian explanation for the origin of light-sensitive cells is discussed in Michael J. Behe, *Darwin's Black Box* (New York: Simon & Schuster, 1996), Chapter 1.

7. Darwin's view, at least when he wrote *The Origin of Species,* seems to have been that God created the universe and the natural laws that govern it, but then turned it loose to run by itself. Unlike many others who hold this view, however, Darwin believed that the law of natural selection cannot produce any determinate outcome, so no specific result of evolution is fore-ordained. As Darwin wrote in a letter to Asa Gray in 1860, he was "inclined to look at everything as resulting from designed laws, with the details, whether good or bad, left to the working out of chance." (Francis Darwin, ed., *The Life and Letters of Charles Darwin* [New York: D. Appleton, 1887], II:105-106) See also Jonathan Wells, *Charles Hodge's Critique of Darwinism: An Historical-Critical Analysis of Concepts Basic to the 19th-Century Debate* (Lewiston, NY: Edwin Mellen Press, 1988).

For more on Kenneth Miller's views, see his book, *Finding Darwin's God* (New York: Cliff Street Books, 2000).

8. The two books in the pile that are not by "strict believers in biblical creation" are *Darwin On Trial,* by Phillip E. Johnson (Washington, DC: Regnery Publishing, 1991), and *Darwinism: The Refutation of a Myth,* by Søren Løvtrup (London: Croom Helm, 1987). Johnson, though a Christian, does not hold to a literal six-day creation or six-thousand year history. Løvtrup is an evolutionary biologist.

Darwinists sometimes claim that it was Darwin who showed that humans are part of nature. But that was never in doubt. Aristotle discusses emotions that humans share with animals in his *History of Animals,* 488b12-20, 508a19-22, 571b9-11, 575a20-32, 581b12-21, 585a3-4, 588a22-31, 608a11-608b18, 629b6-8, and 630b18-23. Catholic theologian and philosopher Thomas Aquinas discusses the animal nature of human beings in the *Summa Theologiae,* First Part ("Treatise on Man") and First Part of the

Second Part ("Treatise on the Divine Government"). In the eighteenth-century, Swedish biologist Carolus Linnaeus classified humans and apes together in one taxonomic family. For Aquinas and Linnaeus, however, the similarities between humans and other animals came from a common creator, not a common ancestor. The issue here is not whether humans have an animal nature, but whether humans are *just* animals.

2

Great Transformations

Fossil whales. Similarities in limb bones. The transition from water to land animals. A "genetic toolkit" common to all animals. The transition from apes to humans.

A. Humans: A Recent Branch on the Tree of Life

This second episode sets out to answer some big questions: "Who are we? Where do we come from? How did we get here? Why do we look the way we do?"

"The story of human evolution," we are told, "is really just a small chapter in a much larger story—the story of all living things." As University of Chicago paleontologist Neil H. Shubin puts it: "Evolution shows us that we're much more connected to the rest of the world, the rest of animal life—than we could ever have imagined."

Accompanied by beautiful photography of wild animals, and shots of scientists chipping away at rocks or peering through microscopes, the narrator continues: "The deeper we dig, the farther back we go, the more we see that everything alive has evolved from a single starting-point. The tree of life has been branching for four billion years, and we can now follow the branches back to their roots."

As the camera focuses on the fossilized fin of an ancient fish lying next to the bones of a human arm, Shubin says: "When we look back over time, we find certain signposts, certain key events—the great transformations, the big evolutionary steps." Then, as a whale rises majestically out of the water, the narrator explains: "Fifty million years ago, land mammals were transformed into sea creatures. Long before that, fish colonized land. At the dawn of animal life itself, the very first bodies appeared. These are just some of the chapters in life's story—our story."

After learning that all of human existence is only a brief moment at the end of a very long history of life, we are told that even though we are latecomers "we have been shaped by the same forces that have shaped all living things. To understand how we fit in, we need to look back to long before our own origins, and see how evolution has shaped other living things."

B. Whale Evolution

The scene shifts to whales gliding effortlessly beneath the waves, and the narrator tells us that their origin "was a mystery." According to University of Michigan paleontologist Philip D. Gingerich, "whales are so different from every other kind of mammal that we can't easily relate them to anything else, and so they're off by themselves as a branch of mammal evolution."

As air-breathers, mammals live mostly on land; but whales and dolphins are mammals that live in water. "But we know that mammals evolved on land," Shubin says, "so it's a real puzzle how whales originally evolved. By understanding how that happens, we'll begin to understand how these big jumps—these big transformations—happen generally."

Gingerich explains how he discovered—and identified—the fossilized bones of a whale-like creature in Pakistan. His search for what scientists call "transitional forms" between land animals and whales later took him to the Sahara Desert, which used to be covered by water. There he discovered numerous fossils of a previously discovered extinct whale, *Basilosaurus.* But Gingerich, unlike those who had gone before him, found tiny leg bones with the fossils—thereby showing that *Basilosaurus* was a whale with legs.

The narrator explains that the land-dwelling ancestors of modern whales might have found food and safety in the water of an ancient sea: "Over millions of years, front legs became fins, rear legs disappeared, bodies lost fur and took on their familiar streamlined shape." The list of transitional forms between ancient land animals and modern whales, we are told, has grown, proving that "the evidence for evolution is all around us, if we choose to look for it."

Ignoring the fact that the transitional series isn't as neat as it is portrayed here, there are at least two problems with interpreting these fossils as evidence for Darwinian evolution. First, it is impossible to determine whether one fossilized species is ancestral to another. According to Henry Gee, chief science writer for *Nature,* "the intervals of time that separate fossils are so huge that we cannot say anything definite about their possible connection through ancestry and descent." The fossils examined in this episode are separated by millions of years and thousands of generations. But it's hard enough to determine who our own great-great-great grandparents are, even though they are of the same species, the time span is measured in hundreds of years, and we have written records to help us. We can only **assume** that these intermediate fossil forms were connected by a chain of ancestry and descent.

Second, mere similarity does not demonstrate an ancestor-descendant relationship. Many of the striking similarities we see today among various organisms were well know to Darwin's predecessors, who attributed them to a common designer. Ford automobiles show a series of transitional forms between the Model T and current models, but all of them were designed and created by intelligent agents. It would make no sense to say that Ford automobiles evolved in a Darwinian fashion unless we could show that a natural mechanism produced them, without any help from human designers. Similarly, before we can call transitional forms between ancient land mammals and modern whales evidence for Darwinian evolution, we must show that a natural mechanism produced them—or at least was capable of producing them.[1]

And that mechanism has to be demonstrated with more plausibility—not to mention evidence—than we see here. To claim merely that "front legs became fins, rear legs disappeared, bodies lost fur and took on their familiar streamlined shape" is not good enough. We have no evidence from modern animals that front legs can become fins, or that a body can assume a radically different shape—much less that a land animal can make the numerous physiological changes it would need for life in the water. Fossils of extinct animals do not necessarily show us descent from a common ancestor, nor do they show us that the change was due to Darwinian natural selection acting on random variations.

The scene changes to an aquarium, where we are told that "bones aren't the only evidence for whale evolution. Their ancestry is also visible in the way they move." The fact that marine mammals propel themselves through the water with up-and-down movements instead of the side-to-side movements characteristic of fish is supposed to indicate their descent from land mammals. But perhaps this is just a common feature of mammals, like air breathing or bearing live young. How does the fact that marine mammals move like other mammals provide evidence for Darwinian evolution?

Neil Shubin returns to conclude the story of whale evolution. "In one sense, evolution didn't invent anything new with whales," he says, "it was just tinkering with land mammals. It's using the old to make the new, and we call that tinkering."

Sort of like what people do with automobiles?

C. Moving onto the Land

Land animals came before whales, but fish came before land animals. So the great transformation that produced land animals preceded the one that produced whales. "It was the moment when fish crawled out of the water."

"The first creatures to leave the water really started something," the narrator explains. "Their ancestors eventually evolved into today's reptiles, birds, and mammals. And these creatures' common origins are still visible in their bodies. Just like us, they all have bodies with four limbs—they're all tetrapods."

Neil Shubin jumps in again: "What that means is that all these different creatures are descended from a common ancestor which had something very similar, or akin, to limbs."

"Just what was that common ancestor," the narrator asks, "and how did it leave the water 370 million years ago?" Shubin and his colleagues find fossils in Pennsylvania suggesting that early tetrapods lived in streams, while Cambridge University paleontologist Jenny Clack finds fossils in Greenland suggesting that fish evolved limbs before they left the water.[2]

Shubin points to the fossil fish fin and human arm skeleton that we saw at the beginning of the episode, and he notes the similarity in the arrangements of their bones. According to the narrator: "With the basic pattern in place, the fin-to-limb transition was merely a series of small changes occurring over millions of years." And a cartoon animation shows us how easy this might have been. But a cartoon animation, no matter how plausible, does not show how real animals in real time could have been transformed from fish into land animals.

Shubin continues: "There's really no goal to evolution. Evolution wasn't trying to make limbs; it wasn't trying to push our distant ancestors out of the water. What was happening was a series of experiments." And the narrator concludes: "Fish experimented with all sorts of survival strategies. . . . The first tetrapods possibly found another way to survive"—by getting out of the water.

So fish "experimented" with survival "strategies" that included growing legs. But clever human biologists have been experimenting with fish for years, and they have not come up with a strategy to make fish grow even the beginnings of legs. What sense does it make to say that fish "experimented" with growing legs?

The truth is that scientists don't know how the first legs—or the first tetrapods, or the first air-breathers, or the first whales—originated. The fossils tell us that aquatic animals preceded land animals, and that land animals preceded whales. On the question of what caused these great transformations, however, the fossils are silent.

But without knowing what caused these great transformations, how can Shubin say with such confidence that evolution had no goal? How does he know?

D. The Cambrian Explosion

As the camera pans over the fossilized remains of ancient animals, the narrator says: "The water-to-land transformation wasn't the first time evolution had experimented with radical new forms of life. An even earlier explosion—perhaps the most significant of all—occurred just over half a billion years ago. This was the one that led to animal life itself."

Shubin elaborates: "Evolution tinkered with fish to make limbs, but fish carry the baggage of their own past. Think of a fish. It has a head, it has a tail, and a bunch of fins in between. That's a body plan—the way the body's put together. But that's just one of many ways of putting animals together." We're shown pictures of jellyfish, a millipede, a beetle, and a crab. "The question is: What sort of tinkering led to these body plans? I mean really what we're dealing with here is the origin of animals."

"According to the fossil record," the narrator says, "animals burst upon the earth rather suddenly," hundreds of millions of years ago. "Scientists call this crucial transformation the Cambrian explosion." Cambridge University paleontologist Simon Conway Morris explains: "The Cambrian explosion was effectively one of the greatest breakthroughs in the history of life. About half a billion years ago, suddenly things change, we have this extraordinary explosion of diversity. And this sudden appearance of the fossils led to this term, the Cambrian explosion. Darwin, as ever, was extremely candid—he said, Look, this is a problem for my theory. How is it that suddenly animals seem to come out of nowhere? And to a certain extent that is still something of a mystery."

It certainly is. In Darwin's theory, all animals are descended from a common ancestor. If the theory were true, we would expect the history of animal life to begin with one form; as time passes, this form would give rise to two or three that are only slightly different from each other; and with more time, these would give rise to other forms, even more different from each other. Finally, after millions of generations, we would see the major differences that now separate clams from starfish and insects from vertebrates.

In the fossil record, however, these major differences appear *first*. In other words, the fossil pattern is exactly the opposite of what we would expect from Darwin's theory. Darwin himself was aware of the problem (as Simon Conway Morris points out); but he hoped that it would be remedied as more fossil discoveries showed the expected long history of gradual divergence before the Cambrian. A century and a half of additional fossil collecting, however, have shown that Darwin was wrong, and the Cambrian explosion was real.[3]

The coverage of the Cambrian explosion in this episode is better than one usually sees in treatments of evolution. Most biology textbooks (including one co-authored by Kenneth Miller, who played a prominent role in Episode One), completely ignore the Cambrian explosion and the challenge it poses to Darwin's theory. *Evolution*'s producers deserve to be commended for including it, though they largely ignore Simon Conway Morris's comment that the Cambrian explosion "is still something of a mystery," and thus pass up a chance to acquaint viewers with the controversy surrounding one of evolution's most exciting unresolved challenges.

Spectacular views of the Canadian Rockies now follow—home of the famous Burgess Shale, which has provided us with much of the best evidence for the Cambrian explosion. After surveying some of the forms found in the Burgess Shale, the narrator says: "All the basic body plans found in nature today are here. Everything that has lived for the last half-billion years came from tinkering with these initial designs. We can even see our own ancestor here." Simon Conway

Morris displays a photograph of a tiny worm, *Pikaia*, and tells us "this might be the precursor of the fish, and so also, I believe—after a long evolutionary story— ourselves."

Neil Shubin returns to summarize what we've seen so far: "So what do we learn by looking at 600 million years of animal history? Evolution's tinkering with mammalness to make whales; in the same way, it's tinkering with fishiness to make tetrapods; and it's tinkering with animalness to make all the different body plans that we see."

There's that "tinkering" again. We've seen that animals burst upon the scene rather suddenly, that land animals came later, and that whales came later still. Things are certainly not what they used to be. But where is the evidence that it all happened through "tinkering"?

Shubin continues: "All these different creatures are variations of the same theme, re-stated over and over again. The question was, What was evolution tinkering with? One of the remarkable discoveries of the last twenty years is that evolution's not tinkering with the bodies, it's tinkering with the recipe, the machinery that builds bodies. What is that recipe? What is that machinery? It's the genes, the genes that build them."

Finally, we are promised some hard evidence of how these great transformations took place. Let's take a look at it.

E. The Genetic Mechanism of Evolution

"Fossils record the changes in animals' bodies over time," the narrator says, "but just *how* bodies changed was unknown. The search for the genetic mechanism of evolution took most of the century. When scientists finally found it, they were astonished by just how simple it was."

In the nineteenth century, geneticist William Bateson had observed that embryos would occasionally develop body parts in the wrong places. Then biologists in the 1940s discovered that they could produce such effects in fruit flies by using radiation. Stanford University developmental biologist Matthew P. Scott explains that, like Bateson, these researchers would occasionally find flies with "one part of the body in the wrong place, or a copy of a normal part of the body in another place."

Watch closely as a fruit fly with an extra pair of wings fills the screen. (Fruit flies normally have two wings, but this one has four.) Note that the wing visible here behind the normal pair is stiff and motionless. That's because the second pair of wings has no muscles. The extra wings are thus useless, and the fly has difficulty flying and mating—though that fact goes unmentioned here. Next we see a mutant fly with no wings (which of course has even more difficulty flying!), then a fly with stubby legs instead of antennae growing out of its head. All three of these mutant flies are cripples, and cannot survive long outside the laboratory.

"The scientists had triggered the changes by damaging the fly's DNA," the narrator says. He continues with an overview of how scientists in the past quarter-century have unraveled some of the ways in which genes affect embryo development. It's a fascinating story about painstaking research, requiring considerable patience and perseverance, which finally ended in success. One of those successes involved deciphering the action of the *Antennapedia* gene, in which mutations can cause flies to sprout legs from their heads. The narrator concludes that the "implications were mind-boggling," because such genes seemed to be "acting like architects of the body."

Could this discovery be generalized to other animals? The scene switches to Switzerland, where University of Basel cell biologist Walter Gehring describes how he removed a gene from a fruit fly that is needed for normal eye development, and inserted a comparable gene from a mouse. Gehring found that the mouse gene did the work of the fly gene. "The fruit fly had grown normal fruit fly eyes," the narrator explains, "using a gene from a mouse. Not only did the two creatures use the same mechanism—they used the very same gene. This was the mechanism behind extra wings, legs sprouting from heads, and Bateson's deformed animals. The century-long search was complete. The genetic engine of evolution turned out to be a tiny handful of powerful genes."

As we watch some more beautiful wildlife photography, University of Wisconsin geneticist Sean B. Carroll interprets the significance of this: "So what this means is—in some ways, some sense—evolution is a simpler process than we first thought. When you think about all of the diversity of forms out there, we first believed that this would involve all sorts of novel creations, starting from scratch, again and again and again. We now understand that, no, that evolution works with packets of information, and uses them in new and different ways and new and different combinations without necessarily having to invent anything fundamentally new, but new combinations."

As a series of brightly-colored cartoons shows the supposedly similar body organization of various kinds of animals, the narrator explains: "Suddenly, the commonality of form among animals was understood. Animals resembled each other because they all used the same set of genes to build their bodies—a set of genes inherited from a common ancestor that lived long ago." Matthew Scott adds: "And what we see now among all the animals are just variations on a body plan that existed half a billion years ago."

"And there's only one inescapable conclusion you can draw from that," says Carroll, "which is: If all of these branches have these genes, then you have to go to the base of that, which is the last common ancestor of all animals, and you deduce it must have had these genes. So the whole radiation of animals, the whole spring of animal diversity has been fed by essentially the same set of genes."

What's wrong with this picture? The story we have just heard ignores two fundamental problems. The first is that (as we saw above) the genetic changes shown here are—without exception!—harmful to the organism. In the wild, all of these changes would be quickly eliminated by natural selection. Geneticists have learned a lot about how genes affect embryo development, but they have not yet found a single mutation that changes the shape of an animal's body in a way that might be useful for evolution outside the laboratory. (Useful mutations occur in some cases of antibiotic resistance, as we shall see in Episode Four; but that's a far cry from changing the shape of an animal's body.)

The second problem is that the "tiny handful of powerful genes" is nowhere near as powerful as we are led to believe. Note that the mouse eye gene inserted into the fruit fly produced a *fruit fly* eye, not a mouse eye. In other words, the gene was not the "architect" of the eye; it merely acted as a switch, enabling the animal to make an eye when and where it needed one. But the gene has nothing to do with the kind of eye the animal makes. It's more like an electrical switch that can turn on a light, a computer, a vacuum cleaner—or whatever else is plugged into it. If these genes are what animals use to "build their bodies," and if all animals have the same set of genes, how come the various kinds of animals are so different from each other? Why don't fruit flies give birth to finches?

In fact, the genes described here are not even turned on by the embryo until the basic body plan is already formed. A fly is a fly, and a mouse is a mouse, long before these genes perform their switching jobs to tell the fly where to grow its legs, or the mouse whether to grow an eye. Whatever it is that builds animal body plans, it is certainly not this "tiny handful of powerful genes."[4]

Once again, we are left without the evidence we were hoping to see. Instead, we are simply assured that evolution is simpler than we thought, and given the same line that modern animals are simply variations of an ancestral body plan that existed long ago.

F. From Ape to Human

"What about us?" the narrator asks. "Our bodies are built from the same genes that build all other animals. Yet we are different. No other animal designs or creates like we do." The camera pans slowly over Michelangelo's Sistine Chapel painting of God touching Adam. The narrator continues: "We seem so special, it's hard not to think that we're somehow an exception to evolution. But of course we're not. The transformation that led to us was no different from the others." In a familiar scene (which is repeated again and again throughout the series), an ape clambers down a log to the ground. "The crucial turning-point seems to have occurred about seven million years ago, when our ancestors left the trees and began to walk on two legs."

According to Arizona State University paleoanthropologist Donald Johanson, this probably first happened in East Africa. He and the narrator explain how walking on two feet seems to have opened the door to the evolution of our brains, though little is known about how our ancestors became bipeds. University of Texas anthropologist Liza J. Shapiro tries to answer this question by studying lemurs, because "we have to know what it was we started from." The narrator explains: "Like lemurs, our early ancestors could move in all sorts of ways." So "they were already adapted to so many movement styles, they could serve as the starting-point for a variety of evolutionary experiments. And that's just what happened."

After being assured that we evolved from a lemur-like animal, and that the striking similarities between chimps and us show that we only recently evolved from a common ancestor, we are told that "the few physical differences that set us apart seem to have made a great difference." Johanson points out some of them on models of human and chimp skeletons, and concludes: "These are minor differences. These are the sorts of tinkering that evolution did to change us into a modern biped." So "what we see is that evolution has worked the same way with us as it has with every single organism on this planet. We're here through a series of chance coincidences, specific adaptations, chosen opportunities."

Words like "tinkering" and "chance" clearly mean something other than what Michelangelo painted on the ceiling of the Sistine Chapel. But—once again—where is the evidence? That we are "built from the same genes that build all other animals"? That lemurs can "move in all sorts of ways"? That there are both similarities and differences between humans and chimps?

The truth is that the evidence for human origins is even weaker than some of the other evidence we've seen. According to Henry Gee, chief science writer for *Nature,* all the evidence for human evolution "between about 10 and 5 million years ago—several thousand generations of living creatures—can be fitted into a small box." Thus the conventional picture of human evolution as lines of ancestry and descent is "a completely human invention created after the fact, shaped to accord with human prejudices." Putting it even more bluntly, Gee concludes: "To take a line of fossils and claim that they represent a lineage is not a scientific hypothesis that can be tested, but an assertion that carries the same validity as a bedtime story—amusing, perhaps even instructive, but not scientific."[5]

So Episode Two, instead of showing us the "underlying evidence" for Darwin's theory, leaves us with a bedtime story.

Notes

1. Not surprisingly, the actual story of whale-like fossils is not as neat as the one told here. There are long-standing disputes over the identity of the land ancestor, the geological position of various fossils, and whether these were the ancestors of modern

whales. Modern molecular studies have added to the controversy. For a short survey of some of the disputes, see Ashby L. Camp, "The Overselling of Whale Evolution," available at:

>   http://www.trueorigins.org/whales.htm#top.

See also:

>   http://www.sciencenews.org/sn_arc98/10_10_98/Fob3.htm

For more on how modern molecular studies have added to the controversy, see Trisha Gura, "Bones, molecules . . . or both?" *Nature* 406 (2000), 230-233; Maureen A. O'Leary, "Parsimony Analysis of Total Evidence from Extinct and Extant Taxa and the Cetacean-Artiodactyl Question (Mammalia, Ungulata)," *Cladistics* 15 (1999), 315-330. See also:

>   http://www.findarticles.com/m1200/19_156/57828404/p1/article.jhtml

On the impossibility of inferring ancestor-descendant relationships from fossils see Henry Gee, *In Search of Deep Time* (New York: The Free Press, 1999). Some passages from Gee's book that deal with human evolution are cited below in the note on human origins.

On the fact that mere similarity is insufficient to establish Darwinian descent with modification, see Jonathan Wells and Paul Nelson, "Homology: A Concept in Crisis," available at:

>   http://www.arn.org/docs/odesign/od182/hobi182.htm

2. For the standard story, see:

>   http://beta.tolweb.org/tree/eukaryotes/animals/chordata/terrestrial_vertebrates.html

Of course, the true story is more complicated than the one presented in this episode; see Michel Laurin, Marc Girondot and Armand de Ricqlès, "Early tetrapod evolution," *Trends in Ecology and Evolution* 15 (2000), 118-123.

The "tinkering" metaphor comes from François Jacob, "Evolution and Tinkering," *Science* 196 (1977), 1161-1166. According to Jacob, an engineer works according to a preconceived plan, uses prepared materials and special machines, and produces things that are as nearly perfect as possible. Evolution, on the other hand, has no plan, works with whatever is at hand, and produces things that are imperfect. But a tinkerer still works according to a plan, though it may be a short-range plan (i.e., to make something useful); and the history of technology is filled with examples of engineered products that were notably imperfect. Most importantly, the sort of creative capacity attributed to natural selection by the tinkering metaphor has never been observed in nature.

3. Darwin wrote in *The Origin of Species* that "if the theory be true, it is indisputable that before the lowest Cambrian stratum was deposited long periods elapsed . . . [in which] the world swarmed with living creatures." Yet he acknowledged that "several of the main divisions of the animal kingdom suddenly appear in the lowest known fossiliferous rocks." Darwin called this a "serious" problem which "at present must remain inexplicable; and may be truly urged as a valid argument against the views here entertained." (Chapter X; page numbers will vary depending on the edition.)

Simon Conway Morris has written about the Burgess Shale in *The Crucible of*

Creation: The Burgess Shale and the Rise of Animals (Oxford: Oxford University Press, 1998). So has Stephen Jay Gould, in *Wonderful Life: The Burgess Shale and the Nature of History* ((New York: W. W. Norton, 1989). See also Simon Conway Morris and H. B. Whittington, "The Animals of the Burgess Shale," *Scientific American* 241 (1979), 122-133; Mark and Dianna McMenamin, *The Emergence of Animals: The Cambrian Breakthrough* (New York: Columbia University Press, 1990); Jeffrey S. Levinton, "The Big Bang of Animal Evolution," *Scientific American* 267 (1992), 84-91; and J. Madeleine Nash, "When Life Exploded," *Time* (December 4, 1995), 66-74.

A fossil bed that documents the Cambrian explosion even better than the Burgess Shale is the Chengjiang, in southern China. The Chengjiang recently yielded fossils of the earliest vertebrates. See Philippe Janvier, "Catching the first fish," *Nature* 402 (1999), 21- 22; D.-G. Shu et al., "Lower Cambrian vertebrates from South China," *Nature* 402 (1999), 42- 46; Jun-Yuan Chen et al., "An early Cambrian craniate-like vertebrate," *Nature* 402 (1999), 518-522.

For a more extended discussion of the challenge posed by the Cambrian explosion to Darwin's theory, see Jonathan Wells, *Icons of Evolution* (Washington, DC: Regnery Publishing, 2000), Chapter 3. In contrast, one biology textbook that covers the topic of evolution but manages to ignore the Cambrian explosion completely is Kenneth R. Miller and Joseph Levine, *Biology* (Upper Saddle River, NJ: Prentice-Hall, 2000).

4. For a detailed discussion of the problems with using four-winged fruit flies as evidence for evolution, see Jonathan Wells, *Icons of Evolution* (Washington, DC: Regnery Publishing, 2000), Chapter 9.

For some general critiques of the idea that "genes build bodies," see H. F. Nijhout, "Metaphors and the Role of Genes in Development," *BioEssays* 12 (1990), 441-446; Brian Goodwin, *How the Leopard Changed Its Spots* (New York: Charles Scribner's Sons, 1994); Steven Rose, *Lifelines* (Oxford: Oxford University Press, 1997); and Jason Scott Robert, "Interpreting the homeobox: metaphors of gene action and activation in development and evolution," *Evolution & Development* 3:4 (2001), 287-295.

5. For the Gee quotations see: Henry Gee, *In Search of Deep Time* (New York: The Free Press, 1999), 23, 32, 113-117, 202.

According to paleoanthropologist Misia Landau, many writings in her field have been "determined as much by traditional narrative frameworks as by material evidence." The typical framework is that of a folktale in which a hero (i.e., our ancestor) leaves a relatively safe haven in the trees, sets out on a dangerous journey, acquires various gifts, survives a series of tests, and is finally transformed into a true human being. When paleoanthropologists want to explain what really happened in human evolution they use four main events. These are: moving from trees to the ground, developing upright posture, acquiring intelligence and language, and developing technology and society. Although Landau found these four elements in all accounts of human evolution, their order varied depending on the viewpoint of the narrator. She concluded that "themes found in recent paleoanthropological writing . . . far exceed what can be inferred from the study of fossils alone and in fact place a heavy burden of interpretation on the fossil record—a burden which is relieved by placing fossils into preexisting narrative structures." *Narratives of Human Evolution* (New Haven, CT: Yale University Press, 1991), ix-x, 148.

In 1997, Arizona State University anthropologist Geoffrey Clark wrote that "we select among alternative sets of research conclusions in accordance with our biases and preconceptions—a process that is, at once, both political and subjective." Clark suggested "that paleoanthropology has the form but not the substance of a science." G. A. Clark and C. M. Willermet (eds.), *Conceptual Issues in Modern Human Origins Research* (New York: Aldine de Gruyter, 1997), 76.

EPISODE

3

Extinction!

Mass extinctions. Dinosaurs and the first mammals. Saving an unspoiled forest near Bangkok. Biological invaders in the Hawaiian Islands. Using a beneficial insect to control weeds in North Dakota.

A. The Permian Extinction

As leaves fall from a tree—presumably a symbol for Darwin's tree of life—University of Washington paleogeologist Peter D. Ward tells us: "Extinction is the termination of a species." At least 95 per cent of all species that have ever lived are now extinct. Extinction is normal, he says, and is happening all the time, at the rate of a few species per year.

We watch various animals foraging for food, and a lioness bringing down her prey. "The extinction of old species that can no longer adapt or compete creates opportunities for new species that can—in an endless cycle," the narrator says. "So evolution and extinction are in balance. But what happens when a planet-wide catastrophe strikes and a great dying begins?"

The scene changes dramatically—to lightning, volcanoes, and fire. Five times in the last half-billion years, we are told, mass extinctions wiped out most species alive at the time. As the smoke clears, we see Peter Ward driving through South Africa to investigate the greatest of these mass extinctions—the one that occurred at the end of the geological period known as the Permian. He stops at an old abandoned farmhouse, and sees from the tombstones in a nearby graveyard that the family that used to live there died within a five-year period about a century ago. "So a hundred years [ago], these people were just wiped off the face of the Earth, and we have no idea what killed them," says Ward. "And if that's the case, how am I going to figure out what killed animals that lived in those hills [gesturing], the fossils of which we have from 250 million years ago?"

In the rocks of those hills, Ward finds evidence that a great catastrophe occurred at the end of the Permian. "So catastrophic was that mass extinction," says Ward, "that even the small creatures have died out. It's not just the mighty, it's the meek." An animation shows us what might—or might not—have caused the Permian extinction. "When species died, they didn't die alone," says the narrator. "The collapse of one helped bring down the others."[1]

Ward explains: "You could almost analogize that to a house of cards. Each species props up another, in a sense." We watch as a huge house of playing-cards teeters in front of us. Ward continues: "Because the creature that you eat is that card that is sitting under you that gives you your energy. Now let's pretend that we start kicking out card after card after card—and that's what a mass extinction does, isn't it? It starts knocking out a species here, it knocks out a species there, but pretty soon lots of species are gone. And it's not just the disappearance of species now—the whole house of cards falls down."

Not everything died in the Permian extinction, however. Ward holds up the skull of a mammal-like reptile. He says that the few lineages that survived the extinction "start evolving, because the world is empty, and empty worlds really begat [a] tremendous amount of evolutionary diversifications."

But how do empty worlds beget new species, exactly? Mass extinction may be an important feature of the history of life; but the question is, how did living things diversify afterwards? *That* is the question Darwin's theory is supposed to answer, but the fact of extinction doesn't help us. Species go extinct, and new ones take their places. This may come as a surprise to people who believe that species never go extinct (if, in fact, there are such people); but how does it provide evidence for Darwinian evolution?

B. Dinosaurs, Mammals and Us

Ward takes us to visit a family that for four generations has been collecting fossils from the early Triassic. He explains that two important groups of land animals diversified after the Permian extinction: dinosaurs, some of which attained enormous size; and small mammal-like reptiles.[2]

The scene shifts to the American Museum of Natural History in New York City, where mammal curator Michael J. Novacek tells us about his childhood fascination with stories of dinosaur fossils found by Roy Chapman Andrews in the Gobi Desert in China. Years later, Novacek went to the Gobi Desert himself—but to look for fossils of early mammals instead.

Novacek explains that early mammals were quite small, dwarfed by the dinosaurs who were their contemporaries. After millions of years another mass extinction occurred, at the end of the Cretaceous period, perhaps caused by a large asteroid that landed in what is now the Gulf of Mexico. This global catastrophe wiped out the dinosaurs—in a chain reaction dramatized once again by a falling house of cards.

The narrator tells us that the mammals survived because they were small and "could take refuge underground." When they re-emerged after the "K-T event" (K stands for Cretaceous and T for Tertiary, the geological period following the catastrophe), the dinosaurs were gone, and the world belonged to them. "Mass extinction made them evolution's big winners—by default."

As we are treated to more beautiful wildlife photography, the narrator tells us that mammals—freed from having to compete with the dinosaurs—"spread out to all parts of the world. They filled every empty niche. They competed, adapted, and diversified, until most of the world's largest animals were mammals. Then, around 5 million years ago, the first human precursors emerged in Africa—mammals that would play an unprecedented role in evolution's future." Bleached bones stick out of the dry soil, and we view a trail of footprints as the narrator explains: "They began to walk upright. They left their footprints in volcanic ash that hardened. They evolved into the genus *Homo*—humankind."

This makes an interesting story. But the K-T event, like the Permian extinction, does not help us to understand evolution. Instead of saying that mass extinction made the survivors "evolution's big winners," the narrator might just as well have said "history's big winners," or "life's big winners." Using the word evolution doesn't add a thing here—except perhaps to give us the impression that an explanation has been provided, when in fact none has.

The fossil skulls we saw certainly provide evidence that there were animals living in the past that are no longer with us, and they suggest that some of those animals had features intermediate between ancient reptiles and modern mammals. As we saw in Episode Two, however, fossils by themselves cannot provide evidence for ancestry and descent, much less evidence that the process occurred through natural selection. Life has a history. But was that history due to Darwinian evolution?

The scene shifts to Bangkok, Thailand. "Today," the narrator continues, "the world is bursting at the seams with people." Bangkok, we are told, has doubled in size in the last two decades, and now has more than 10 million inhabitants. In the past ten thousand years, we have out-competed—and thus driven extinct—many other species. The narrator concludes: "We may be the 'asteroid' that brings on the sixth great mass extinction."

C. Studying a Remote Forest

We fly over a mist-shrouded valley as the narrator describes a national park to the west of Bangkok, where the human population is nil. Wildlife Conservation Society scientist Alan Rabinowitz came here to study the ecosystem.

Rabinowitz says: "We're in grave danger of the 'empty forest syndrome'—having a beautiful, seemingly intact forest on the surface, but inside that forest the natural components which maintain the flow of energy through the system—which has developed through millions of years of evolution—it's disrupted. Now people say, So what does it matter if one component's gone? What if you don't have the Sumatran rhino? What if the civet species are all gone—or other things?

But each thing has evolved to play an incredibly important role within this complex puzzle."

The narrator explains that Rabinowitz is here to find out if the forest still has a balanced ecosystem in which evolution can continue without being affected by the increasing rates of extinction elsewhere. He is especially interested in large carnivores. Since they depend on species below them in the energy chain, they're the first to go when the ecological house of cards begins to collapse.

Rabinowitz uses remote cameras triggered by motion sensors to "catch" nocturnal animals on film that otherwise would be almost impossible to find. Finding that some of his cameras have been stolen or vandalized, Rabinowitz fears that humans are encroaching destructively on this remote forest.[3]

"There's no doubt that the major cause of extinction on a global level is human-related," he says. "Everything from clear-cutting forests, to removing intact habitats, to just desecrating them, changing them." The narrator adds: "Habitat destruction is the number one cause of extinction."

D. Biological Invaders

The scene shifts to the Hawaiian Islands, as the narrator explains how they were formed from undersea volcanoes and then colonized by living things from distant lands. "Thousands of species made it by sea or by air. They evolved until they were found nowhere else on earth."

Fordham University paleoecologist David A. Burney and his son explore the Hawaiian Islands "to better understand what happened after the Polynesians arrived" centuries ago. We watch as Burney drains a sinkhole, "revealing ten thousand years of sediment and a story of evolution that's just as long."

Burney finds evidence that the Polynesians brought stowaway rats and a few other non-native species with them. But the Europeans brought many more. "We're now to the point," Burney says, "where there are about a thousand native species of plants in the Hawaiian Islands and over a thousand naturalized invasive species—things that have been introduced by people. Evolution has now entered a new mode. Something new altogether is happening, and it has to do with what humans do to the evolutionary process."

The scene shifts to various modes of air and sea transport, as the narrator explains how hitchhiking species colonize virgin territories and become pests. A Hawaiian agricultural inspector talks about the difficulty of keeping biological invaders out of the islands, and Burney predicts that biological invasions will be visible in the fossil record a million years from now.

"Scientists have a term for biological invaders," the narrator tells us. "They call them 'weed species.' Like weeds, they survive and adapt almost anywhere, and push out the native species—sometimes to the point of extinction. They are the ultimate survivors."

After Novacek returns to speculate briefly on why biological invaders tend to be so successful, the narrator continues: "Of all the weed species on earth, we [human beings] are the most mobile, the most adaptive, and the most flexible—by far. The good news is, we'll probably be around for a long time. The bad news is, the world around us may be very different. Every species we drive towards extinction, no matter how inadvertently, is one less species that might help prop up the others."

Re-enter Peter Ward, who ties this thread back into the main theme of the episode: "So the question is, in our own modern world, with our own house of cards, How close are we to that whole edifice coming down? Have we reached that threshold?"

Curiously, there are several unanswered questions in what we have just been told. First, what, exactly, favors evolution? Does evolution proceed mainly when there is intense competition for survival, as Darwin proposed? Or does it proceed mainly when competition is eliminated through mass extinction, or through migration to an uninhabited island? Maybe it's both. Or maybe it's neither.

Second, is extinction good or bad? From an evolutionary perspective, it seems, extinction is a good thing, since it provides opportunities for surviving organisms to diversify. From a human perspective, however, our own extinction would be a bad thing, so we should interfere with evolution to preserve ourselves. Why, then, do we need an evolutionary perspective?

Third, What does evolution have to do with how some species replace other species over a span of ten thousand years? None of the species are changing. We saw more evolution in bacteria that develop resistance to antibiotics—and even that didn't really help Darwin's theory about the *origin* of species.

E. Leafy Spurge

The scene shifts to the rolling grasslands of North Dakota. "It's where one of the battles against human-caused extinction is being fought," the narrator says. "The enemy here is a weed called 'leafy spurge'—so adaptive and tenacious, it threatens to kill off all the native grasses. It's already spread across a million acres. Accidentally brought by pioneers in bags of seeds a century ago, the settlers' descendants are now faced with the consequences."

A rancher explains that leafy spurge limits the number of cattle he can put in a pasture, because the cattle won't eat the grass if it is infested with the weed. "I look at it as cancer to the land," the rancher says, "and it makes the land just totally useless." There is a chemical available that kills leafy spurge, but it is prohibitively expensive.

The narrator continues: "If the chemical won't stop it, what's left? How can the farmers fight back against a super-adaptive invader that threatens to drive native species to the brink of extinction, and take over their ecological niche? The solution may be another invader—discovered when scientists learned what kept leafy spurge in check in its native Russia. It's the flea beetle—a case of fighting evolutionary fire with fire." Flea beetles eat leafy spurge, and thereby help to keep it under control.

We watch as North Dakota ranchers spread flea beetles throughout their pasturelands in an effort to control leafy spurge. Then the narrator says: "So now we're in a race most of us don't even know we're running—to learn as much as we can about the meaning and message of extinction before it's too late."

But extinction has nothing to do with the leafy spurge story. Although the weed is out-competing native grasses in some areas, it is an exaggeration to say that it "threatens to drive native species to the brink of extinction." Furthermore, it does not render land "totally useless," except to cattle ranchers. Although cattle are repelled by leafy spurge, sheep and goats are not; the latter even seek it out.

The producers of *Evolution* want us to think that a "grounding in evolution is key to our understanding" of agriculture. Perhaps that's why they included the leafy spurge story in their PBS series. But the use of flea beetles to control leafy spurge is not "fighting evolutionary fire with fire"—it's just a variation on an ancient agricultural practice. Farmers have been using one organism to control another for centuries. For example, ants were used in ancient China and Yemen to control pests that would otherwise destroy citrus groves. The practice is known as "biological control," and it does not depend on a "grounding in evolution."[4]

F. What's Evolution Got To Do With It?

The scene then shifts back to Alan Rabinowitz in Thailand. Happily, the other teams of scientists working with him had more success with their cameras than he had. Their films show that the forest is still populated with large carnivores, and thus still has a healthy ecosystem.

Rabinowitz reflects on the implications of this: "There are still places left where the natural evolutionary processes are going on. Most of my career involves documenting extinction, or species on the verge of extinction. But every now and then, you get a place like this. And you say, It's not lost yet. It's not gone yet. Knowledge is definitely our greatest tool against extinction—there is no, no doubt about it. Without knowledge, we continue in the dark. Many species are on a very quick downward slide, possibly to extinction, faster than they would be normally, because of human-related activities. But we're not at an end-point here, by any means. We're still in the middle of a completely complex, changing scenario. Evolution is going on around us."

The "evolution" that Rabinowitz sees going on around us, however, is not "evolution" in Darwin's sense. The "evolution" to which Rabinowitz refers involves the displacement of some species by others, but the "evolution" that matters in Darwin's theory is the origin of new species. And there is no evidence here for the latter.

No one would dispute Rabinowitz's claim that knowledge is important, and many people would approve of his efforts to preserve the unspoiled beauty of remote forests. But his comments about evolution are superfluous. Scientists do not need evolution to engage in conservation efforts, and it is questionable whether they even need it to understand ecology. According to evolutionary biologist Peter Grant, past president of the American Society of Naturalists, "an ecologist's world can make perfect sense, in the short term at least, in the absence of evolutionary considerations."[5]

The episode closes with a reading from Darwin's *The Origin of Species*: "We need not marvel at extinction. If we must marvel, let it be at our own presumption in imagining for a moment that we understand the many complex contingencies on which the existence of each species depends. The appearance of new species and old species are bound together."

What does this last statement mean? Species go extinct, and new ones appear to take their places. But we still have seen no evidence of one species changing into another—much less through natural selection, as Darwin claimed. Extinction, as Peter Ward explained, is the termination of a species, not its transformation.

Extinction happens. But what's evolution got to do with it?

Notes

1. For more information on mass extinctions, see Peter D. Ward, *Rivers in Time: The Search for Clues to Earth's Mass Extinctions* (New York: Columbia University Press, 2001)

    Based on new fossil discoveries, some scientists are now questioning whether the "Big Five" mass extinctions were really as big as previously believed. See Richard A. Kerr, "Mass Extinctions Facing Downsizing, Extinction," *Science* 293 (2001), 1037.

2. Although not much is said about them here, mammal-like reptiles are often cited as good evidence for Darwinian evolution. Not surprisingly, however, the story is more complicated than promoters of Darwinian evolution make it out to be. For one review, see Ashby L. Camp, "Reappraising the Crown Jewel," at:

    http://www.trueorigins.org/therapsd.htm

3. Alan Rabinowitz's work in Thailand and neighboring Myanmar was done under the auspices of the World Conservation Society, which is affiliated with the Bronx Zoo in New York. For more information, see:

    http://wcs.org/home/wild/Asia/tiger

    http://wcs.org/home/wild/Asia/2688

4. Leafy spurge produces an irritating sap that is harmful and distasteful to cattle, so it renders grazing land unsuitable for cattle. As this episode mentions, flea beetles are one way to control leafy spurge. Other ways to control it without resorting to chemical

pesticides include grazing with sheep or goats (goats actually seek it out), or planting wheat grass or wild rye. For general information about leafy spurge, go to:

http://users.aol.com/prideedu/leafy.htm.

For more information about the biological control of leafy spurge, go to:

http://www.ext.nodak.edu/extpubs/plantsci/weeds/w866w.htm

http://www.ext.nodak.edu/extpubs/plantsci/weeds/w1183w.htm

Although the use of flea beetles (as described in this episode) did not start until the 1980s, there is nothing new about biological control. In China and Yemen, farmers have been using ants to control crop pests for centuries. In India, intercropping (planting mixed crops to combat pests) has been used for centuries. Such ancient practices obviously owe nothing to Darwin or his theory. Here are some quotes from web sites of interest:

> Biological pest control was used by the ancient Chinese, who used predacious ants to control plant-eating insects. In 1776, predators were recommended for the control of bedbugs.

http://www.comptons.com/encyclopedia/ARTICLES/0125/01429248_A.html

> The Chinese were far ahead of the Western world in natural pest control. In the countryside frogs were always a forbidden food because they ate insects. Praying mantises were released in gardens among the chrysanthemums to drive away leaf-eating insects. The most remarkable and economically important of the ancient Chinese biological weapons was the yellow citrus killer-ant. Its use is described in Hsi Han's *Records of the Plants and Trees of the Southern Regions,* written in A.D. 340: "The Mandarin Orange is a kind of orange with an exceptionally sweet and delicious taste. . . . The people of Chiao-Chih sell in their markets [carnivorous] ants in bags of rush matting. The nests are like silk. The bags are all attached to twigs and leaves, which, with the ants inside the nests, are for sale. These ants do not eat the oranges, but attack and kill the insects which do. In the south, if the mandarin orange trees do not have this kind of ant, the fruits will be damaged by the many harmful insects, and not a single fruit will be perfect.

http://www.inventions.org/culture/ancient/pest.html

> Yemenis were among the earlier nations that used biocontrol agents for the control of agricultural pests. . . . In traditional agriculture farmers in Tihama, Taiz and Hadramout used to collect predatory ants (qu'as) from mountains to control date palm pests which attack fruits. This has been described in a 13th century agricultural text by al-Malik al-Ashraf 'Umar of Rasulid, Yemen. . . . When we asked old farmers about this practice they confirm that. They added that they have to put some sticks to make ants to travel from one tree to another.

http://www.aiys.org/webdate/pelbaa.html

> Intercropping in India: Farmers in the developing world have been growing two or more crops together on the same piece of land for many centuries. In India, as many as 84 different crops are used in mixed cropping, but seldom do we find more than four at a time, and a relatively simple mixture of only two or three crops is most common. . . . Although research on intercropping may have started to provide an understanding of why the farmer used such mixtures, and to help improve his

productivity in ways relevant to his practice, it has now been shown that intercropping may have several advantages over sole cropping. It appears to make better use of the natural resources of sunlight, land, and water. It may have some beneficial effects on pest and disease problems, although the overall results are somewhat inconclusive.

http://ourworld.compuserve.com/homepages/rbmatthews/rbm_ic1.htm

5. Peter Grant's comment about ecology is from "What Does It Mean to Be A Naturalist at the End of the Twentieth Century?" *The American Naturalist* 155 (2000), 1-12, 9.

4

The Evolutionary Arms Race

Multi-drug resistant tuberculosis in Russia. Lessons from a cholera epidemic. Feline immunodeficiency virus. Leaf-cutter ants and symbiosis. Allergies and the importance of interactions among species.

A. Microbes as Predators

"The Cold War is history," the narrator begins, "but Russia is in the grip of an arms race—an evolutionary arms race, with an enemy the naked eye cannot see." After some scenes of predators catching their prey, and bacteria multiplying under a microscope, we find ourselves in a room crowded with prison inmates—some of them coughing. "A deadly microbe is evolving in Russia's prisons," the narrator continues ominously, "consuming the bodies of men. As it escapes prison walls, it attacks new prey, without preference, without warning." Suddenly we are transported to New York City. "Now the killer is spreading beyond Russia, and no one is immune. Unseen, the microbe is evolving into mutants that may soon elude our best defenses. Will we lose this arms race, or can we reach an evolutionary truce with a mortal enemy?"

The scene switches to western Oregon, "home to one of evolution's most extreme and deadly creations"—a species of newt that defends itself from predatory snakes with tetrodotoxin, a potent nerve poison. Each newt produces enough toxin to kill scores of other animals, but why so much? Research shows that the garter snakes that prey on them have a certain amount of resistance to the toxin. A newt that produces slightly more toxin than its neighbors might overcome the snakes' resistance and survive. If its offspring inherit the ability to produce more toxin, subsequent generations might evolve higher toxin levels. The snakes, in like manner, might evolve higher levels of resistance, and the result would be an "evolutionary arms race" between the two species.

Evolution, we are told, is driven not just by physical forces such as climatic change, but even more by biological forces—the ways species interact with each other. As we watch some more wildlife photography, the narrator asks: "What made the lion fast, and the zebra fierce? What drove the development of tooth and claw? The deadly dance of predator and prey has shaped the evolution of countless species. There may have been a time on an ancient savanna when hungry beasts hunted our ancestors, and drove the evolution of our own species."

We find ourselves once again on crowded city streets, as the narrator continues: "But since the dawn of civilization, only one kind of predator has truly threatened us. The microorganisms that cause disease consume us from the inside out." They also reproduce much faster than we do—a fact dramatized by time-lapse microphotography. "By evolving much faster than we do, microbes have eluded the body's defenses, and left their mark on our history."

The bacteria that cause tuberculosis have been detected in Egyptian mummies, and have preyed on people for thousands of years. A different microbe caused the "Black Death" in the fourteenth century, which killed a third of all Europeans. And still another caused the 1918 flu epidemic, which claimed 20 million lives. "We were virtually defenseless against these microscopic killers until recently," the narrator says. An old newsreel shows a hospital—"a battlefield in man's total war against disease"—and calls antibiotics "the miracle drugs of our time." Antibiotics initially seemed to be so successful that by 1969 the U. S. Surgeon General thought the war on infectious disease had been won. But he spoke too soon.

B. Drug-resistant Tuberculosis

We return to Russia, to the room crowded with prison inmates. Since the fall of the Soviet Union, the narrator says, Russia's prison population has soared. "But overcrowding, poor nutrition, and scant sanitation are not the worst of the prisoners' worries. Now tuberculosis stalks these men." Microbes that might lie dormant in otherwise healthy people erupt into active disease in these men, because their immune systems have been weakened by unhealthy lifestyles and prison overcrowding.

Many of these victims were previously treated for tuberculosis (TB), but their treatment was not continued long enough to cure them completely. Describing one such victim, the narrator says: "Evolution has occurred inside his body." When he was first diagnosed with TB, he was given drugs that killed some bacteria but spared "the ones with mutations that made them resistant to the drugs. As these survivors multiplied, they passed along their protective mutations to all their descendants. In this way, the bacteria evolved into a new drug-resistant strain."

In fact, he and more than 30,000 other Russian prison inmates have TB that is resistant to more than one drug. Although there are now new antibiotics to treat strains with multi-drug resistance, they are expensive and hard to get. When the prisoners are released, they can spread these resistant strains to the general population—and thus to the rest of the world, including the U.S.

Doctors and nurses scurry around an emergency room to dramatize the possible consequences of a TB epidemic in a U.S. city. The scene is frightening. "And TB is just the tip of the iceberg," says the narrator. "The microbes that cause malaria, pneumonia, gonorrhea, and scores of other infectious diseases are evolving drug resistance."

"We've created this problem," a researcher tells us. "Multi-drug resistance is a man-made problem." This is because antibiotics are being used too much. "By developing as many antibiotics as we have over the last fifty years, we've essentially accelerated an evolutionary process. The outcome is that we're going to have *more* drug-resistant microbes—to the point where some of the most dangerous bacteria will not be treatable. We're racing against the microbe every day, and unfortunately we're losing."[1]

There may be a solution, however. "In an arms race without end, the more drugs we launch at microbes the more resistance they evolve," the narrator says. "It may be time to change our strategy, and make evolution work *for* us."

C. Can Cholera Be Domesticated?

The camera pans over a tree-covered college campus, and the narrator continues: "If we can make microbes more resistant, then we can also make them less harmful to us—less virulent."

Amherst College evolutionary biologist Paul Ewald explains: "When people are looking at the antibiotic resistance problem, they see evolution as sort of the bad guy. It's the evolutionary process that's led to antibiotic resistance—and that's true. But just as easily, we can have evolution being the solution. In other words, we can have evolutionary processes leading to organisms becoming more mild."

According to Ewald, microbes that are transmitted through direct person-to-person contact—such as cold viruses—tend to be mild, because they require basically healthy people for their transmission. But microbes that are transmitted through insects, food, or water—such as cholera—tend to make people very sick. "Once we understand the factors that favor increased harmfulness and decreased harmfulness," Ewald reasons, "then we can look at all of the things we do in society. We can ask the question, 'Are we doing certain things, or can we do certain things, that would favor organisms evolving towards mildness.'"

Ewald studied a 1991 cholera outbreak in South America that sickened over a million people and killed almost 11,000, in order to "document evolution in action." Ewald explains: "If you have contaminated water allowing transmission, we expect the cholera organism to evolve to a particularly high level of harmfulness. And that's exactly what we see. We find that bacteria that had invaded countries with poor water supplies evolved increased harmfulness over time."

"If, instead, we clean up the water supplies," Ewald continues, "then we force the bacteria to be transmitted only by routes that require healthy people. And what we find is that when cholera invaded countries with clean water supplies, the organism dropped in its harmfulness. Those bacteria evolved [a] lower level of toxin production—they actually became more mild through time.

"People would still be getting infected, but the infections would be so mild that most people wouldn't even be sick. So the cholera outbreak in Latin America suggests that we may need only a few years to change the cholera organism from one that would often kill people to one that hardly ever causes the disease. What we're suggesting here is that we can domesticate these disease organisms."

But there are serious problems with Ewald's story. First, he claims that microbes spread through person-to-person contact tend to be less harmful than those spread by insects, food and water. But TB—the harmfulness of which was just impressed on us—spreads through person-to-person contact. So did the 1918 flu, which killed more people in less time than the infamous Black Death. Since cholera is transmitted through water or food—not through person-to-person contact—regardless of whether it is mild or harmful, Ewald's hypothesis is wrong from the start.

Second, the solution Ewald proposes—to clean up water supplies—owes nothing to Darwinian evolution. Clean water prevents cholera epidemics primarily because it prevents the bacteria from spreading, not because it spreads bacteria that have evolved to become "more mild." And before Darwin even published his theory, deaths from infectious diseases in England had already declined dramatically because of the general improvement in sanitation and nutrition during the eighteenth and nineteenth centuries.[2]

The lesson to be learned from this is that our first lines of defense against infectious diseases are to keep our food and water clean, and to maintain our general health by eating nutritious foods. Although no one who has ever needed antibiotics will dispute their usefulness, their contribution to public health is minor compared to improved sanitation and nutrition.

So if you don't want to die of cholera, drink clean water and eat clean, nutritious food. Evolution has nothing to do with it.

Like all other infectious diseases, TB declined dramatically in England before the advent of modern medicine, and for the same reasons. As we were told a few minutes ago, Russian prison inmates develop active TB because their immune systems have been weakened by unhealthy lifestyles and poor nutrition. Add overcrowding and poor sanitation, and illness is to be expected. Of course, the problem of multi-drug resistance cannot be ignored; but once again, general sanitation and nutrition turn out to be far more important than evolutionary considerations.

In any case, evolution in cholera virulence and TB antibiotic resistance involves only changes within existing species—just as we saw in the case of HIV. The TB we battle today is the same species as the TB found in Egyptian mummies.

We didn't need Darwin to teach us about minor changes within existing species; in the form of domestic breeding, such changes have been understood and used for centuries. True, Darwin realized that a similar selection process operates in the wild. But where is the evidence for his theory that this process explains the origin of *new* species—in fact, of *every* species?

If we were going to find evidence for Darwin's theory anywhere, we would expect to find it in bacteria. Many species of bacteria reproduce several times an hour, so scientists can study thousands of generations in a single year. Bacteria have been experimentally subjected to intense selection and potent mutation-causing agents for decades, yet no new species have emerged. As British bacteriologist Alan H. Linton wrote just recently: "Throughout 150 years of the science of bacteriology, there is no evidence that one species of bacteria has changed into another." But none of this information is shared with the viewers of the PBS *Evolution* series.[3]

D. Feline Immunodeficiency Virus

We leave the Amherst College campus and go to the National Zoo in Washington, DC, as the narrator says: "Peaceful co-existence with disease organisms may seem like a revolutionary idea. But throughout evolution, other species have also reached a truce with these microscopic killers."

Geneticist Stephen J. O'Brien describes his study of feline immunodeficiency virus (FIV), which is associated with an immune-deficiency disease in domestic cats. He examined DNA from every species of cat—from cheetahs to ocelots, and lynxes to lions—and concluded that "virtually every species of cats on the planet had been exposed to and infected with a version of feline immunodeficiency virus." But no species of wild cats suffer from immune-deficiency disease. Apparently, they are immune to the effects of FIV.

Applying evolutionary thinking, O'Brien speculates that FIV first infected the cats' ancestors three to six million years ago. The narrator describes the supposed consequences: "It decimated the animals. But a lucky few were born with mutations that happened to make them immune to the virus. These survivors passed on their protective genes to most cat species alive today. Domestic cats are a young species, and a recent conquest for FIV. But wild cats have reached the end of a long phase of adapting to a once-deadly foe."

"Sadly," the narrator continues, "we have just begun such a phase with a close relative of FIV—the human immunodeficiency virus. But the example of the cats convinced O'Brien that some people must be endowed with genetic mutations that make them immune to HIV. He set out to find them." And O'Brien found a mutation that he concluded helps protect some people from HIV infection. Applying evolutionary thinking again, O'Brien speculates that this mutation may also have protected Europeans from the Black Death in the fourteenth century.

There are problems with the FIV story, however. First, if all modern members of the cat family inherited a mutation from their common ancestor that protects them from the effects of FIV infection, and domestic cats are the youngest members of the family, then why didn't they inherit the mutation, too? Perhaps the immunity of wild cats is not due to a mutation at all, and it is the environment of domestic cats that renders them susceptible to disease. But this possibility is not even mentioned.

Second, and more importantly, nothing in this story helps us to understand how FIV or HIV—much less cats or humans—evolved. As we have already seen, changes in an existing species are a far cry from the origin of a new species. Even if everything we have just been told about FIV and HIV were true, it would not provide evidence that "leopards and lichens, minnows and whales, flowering plants and flatworms, apes and human beings" evolved from a common ancestor through natural selection and random variation, as Darwin's theory implies.

Third, nothing in this story shows that an understanding of evolution is necessary to medical research. O'Brien's speculations about the evolutionary history of FIV have done nothing to cure domestic cats of their disease. And his finding that some people have a mutation that supposedly protects them from AIDS has done nothing for the victims of that disease. The producers of *Evolution* claim that evolution "touches our daily lives in extraordinary ways," especially when it comes to medicine. Yet this story shows nothing like that.

E. Symbiosis and Leaf-Cutter Ants

"But evolution arises not just from conflict and competition," the narrator says, "the history of life is also the story of different species joining forces."

There is another force, we are told, that is just as important as competition in building up "the magnificent super-structure" of the world of living things, and that is "cooperation—what we call symbiosis, and particularly mutualistic symbiosis. That is intimate living-together of different kinds of organisms, in which there is a partnership which benefits both of the partners."

After being shown several examples of this, we are introduced to leaf-cutter ants in the Amazonian jungle. These ants are like gardeners—they harvest leaves that they chew into a pulp, then they use the pulp to grow a fungus that supplies them with sugar. This is mutualistic symbiosis: The fungus depends on the pulp provided by the ants, and the ants depend on the sugar produced by the fungus.

Remarkably, the ant's "gardens" are pest-free, unlike human gardens. It turns out that the many of the ant-gardeners are covered with a white, waxy coating of

Streptomyces bacteria. Streptomyces—the source of medically useful antibiotics such as streptomycin—keeps in check an aggressive mold that would otherwise devastate the cultivated fungus.

"The ants have been using antibiotics to kill the mold in their gardens for some fifty million years," says the narrator, "so why hasn't the mold evolved antibiotic resistance?" The answer, we are told, is that the Streptomyces bacteria covering the ants is probably evolving along with the aggressive mold that it keeps in check. The result is supposedly "an evolutionary arms race that has continued for fifty million years"—though we are not shown any evidence for this at all.

So leaf-cutter ants provide us with an excellent example of mutualistic symbiosis, and may also provide us with another example of an evolutionary arms race. This is a fascinating story. But what does it have to do with Darwinian evolution? We have seen that several species can co-exist in a symbiotic relationship. But we have not seen evidence for how that relationship developed, much less for how the species originated in the first place.

F. Microbes Can Be Good For Us

Microbes are an important part of our world, we are told, yet we seem to do everything in our power to avoid contact with them. "Is it possible we're making our world *too* clean?" asks the narrator.

We visit a German pediatrician who treats allergies and asthma—"disorders in which the immune system overreacts to harmless substances," explains the narrator. Research conducted by the pediatrician suggests that children who live in villages suffer more from such disorders than children who live on nearby farms. She finds that children who come into regular contact with livestock are less likely to develop allergies and asthma.

The pediatrician suspects that high levels of microorganisms in the stables help children to form healthier immune systems. "Microbes have been around us always," she concludes, "and probably we need to find the balance between eradicating the harmful effect of bacteria, and maybe also taking the beneficial components of this."

So we need to drink clean water, and eat clean food; but perhaps we don't want our environment to be *too* clean.

As the episode closes, we are told that it's a big mistake to separate ourselves too much from the rest of living things. After all, most species are our friends rather than our enemies. In fact they are essential to our existence, and we would do well to learn more about them. Surely, these are wise words. But do we really need to understand Darwinian evolution in order to appreciate them?

In fact, most of what we've heard in this episode would have made just as much sense if the word "evolution" hadn't been used at all. True, we can use the word to explain why newts make an excessive amount of poison; but we haven't thereby explained how newts acquired the ability to make the poison, much less how newts originated in the first place. Minor changes within species have been known for centuries. What we need to see is evidence that the same process can produce the big changes required by Darwin's theory, but that evidence has not been forthcoming.

Nor does this episode show that we need to understand evolution in order to practice good medicine. True, we can use the word to describe how TB becomes resistant to antibiotics; but it's still the same TB we find in Egyptian mummies, and a healthy lifestyle may still be the best defense against it. Cholera epidemics can be prevented without knowing anything at all about evolution, and studying FIV evolution has done nothing to help AIDS sufferers.

Clearly, the "evolutionary arms race" metaphor is not the only way—and may not even be the best way—to understand our relationship to bacteria. Although this episode started with ominous warnings of a microbial holocaust, it ends with the comforting assurance that most microbes are our friends. Darwinism seems to explain everything and its opposite equally well. But if competition and cooperation are equally compatible with evolution, what difference does Darwinism really make?

Notes

1. Multi-drug-resistant TB is an important public health problem. For more information, go to:

      http://www.biomedcentral.com/news/20010402/05 or

      http://www.who.int/gtb/publications/dritw/index.htm or

      http://www.hopkins-tb.org/news/

2. For more information on Ewald's hypothesis, see Paul W. Ewald, *Evolution of Infectious Diseases* (Oxford: Oxford University Press, 1996). For more information about cholera, go to:

      http://www.bact.wisc.edu/Bact330/lecturecholera or

      http://vm.cfsan.fda.gov/~MOW/chap7.html

    The decline in infectious diseases (including TB) from the seventeenth century to the nineteenth largely preceded the advent of modern medicine, and was due to general improvements in sanitation and nutrition. See Thomas McKeown, *The Role of Medicine* (Princeton: Princeton University Press, 1979).

3. Alan H. Linton is emeritus professor of bacteriology at the University of Bristol (U.K.). The quotation is from *The Times Higher Education Supplement* (April 20, 2001), 29.

    The person who has probably come closer than anyone to producing new species of bacteria is Michigan State University biologist Richard E. Lenski. But Lenski has only been able to produce "an incipient genetic barrier between formerly identical lines"—a barrier which he admits is "much smaller than the barrier between such clearly distinct species as *E. coli* and *Salmonella enterica.*" See Martin Vulic, Richard E. Lenski, and

Miroslav Radman, "Mutation, recombination, and incipient speciation of bacteria in the laboratory," *Proceedings of the National Academy of Sciences USA* 96 (1999), 7348-7351.

EPISODE

5

Why Sex?

Genetic variability as a reason for sexual reproduction. Sexual selection and peacocks' tails. Lessons from chimpanzees and bonobos. Evolutionary psychology on the role of sex in brain evolution.

A. Why Sex?

A male peacock struts in front of the camera. "Peahens only mate with well-endowed males," the narrator says. "No fancy display? No sex. No passing of genes to the next generation. That's something every living thing is programmed to do."

The scene shifts to fighting baboons, and the narrator continues: "It's worth fighting for, maybe even dying for. In fact, from an evolutionary perspective, sex is more important than life itself." We watch salmon spawning and dying, then a human family appears. "While we won't trade our lives for sex, most of us will risk death to protect our children, the carriers of our genes. Evolution is a story written over countless generations. To inherit and pass on genes is to be part of that story."

We meet Rutgers University evolutionary geneticist Robert C. Vrijenhoek, who tells us: "That's our immortality. That's what connects us to humans on into the future. That's what's connected us to all of our ancestors in the past. That's what connects us to the ancestors that were fish, the ancestors that were protozoans, and the ancestors that were bacteria. It's the single thread that connects all of life on this planet."

"Sex and genes," the narrator says, "driving behavior, driving evolution."

In southwest Texas, a research crew is rounding up parthenogenetic lizards—a species that consists entirely of females, and does not require males to reproduce. The crew leader says: "Some people think they actually have to have some kind of lesbian behavior where a female mounts a female to get the eggs to develop." We watch as one lizard performs erotic movements on top of another. "That hasn't been really proven yet, but it's an interesting hypothesis." The existence of this all-female species, however, raises a fundamental question: "Why is there sex? I mean, are males really necessary?"

Accompanied by some old movie footage and some wildlife photography, the narrator explains the supposed disadvantages of relying on males to help pass on genes, but notes that almost every life on the planet is nevertheless a result of sexual reproduction. The scene shifts to a husband and wife and their adopted baby, and the narrator concludes: "So males must play a critical role, and sex must offer an advantage. Whatever it is, it's buried deep within us."

We are told that the "biological imperative, as we all know, is to pass on genes." Since sexual reproduction is so common, "there has to be some fundamental biological, evolutionary reason for sexual reproduction. This has been one of the major questions in biology for a very long time."

The emphasis on sex in these opening scenes is striking. Sex is more important than life itself. Sex is our immortality. Even the story of parthenogenetic lizards— for whom sex is irrelevant—is illustrated with a lesbian sex scene. This episode is sure to be an attention-getter. But how much of it will it be science?

B. The Biological Reason for Sex

We accompany Vrijenhoek as he conducts research on minnows that inhabit small streams in Sonora, Mexico. The narrator explains that he "hopes to find clues to the enduring mystery—Why sex?"—by studying a species of fish that includes both sexual and asexual reproducers. According to Vrijenhoek, the sexually reproducing fish are more resistant to infestation with parasites. His hypothesis is that this is due to the greater genetic variability that comes from exchanging genes with other members of the population. Sexual reproduction, he believes, leads to new combinations of genes that confer protection against parasites.

"Here was a solution to the mystery of sex," the narrator says. "It's the best defense against rapidly evolving enemies. Or so it seemed, until a bad drought dried up the pools, killing the minnows and throwing everything into question." After the minnows returned, the parasites were decimating the sexual fish, and the asexual ones were doing quite well. But an experiment convinced Vrijenhoek that this was only because the drought had led to inbreeding among the few survivors, and had thus reduced the genetic variability of the sexual population. He concludes that his hypothesis remains intact, and that the principal benefit of sexual reproduction is that it increases genetic variability.

"That's what sex does," Vrijenhoek says. "Sex generates variability among offspring. And when you take that away from a sexual reproducer by inbreeding them, cloning them, you've lost the very benefit of sex. It's that generation of an immense amount of diversity, that diversity of your offspring, that provides challenges to everything around it: challenges to the parasites, challenges to the viruses, challenges to your competitors. That's the beauty of the sexual process— [it] is the variation and wonderful diversity it creates."

As we watch human families and children, the narrator repeats the point: "Sex generates variation, which improves a species' chances of survival in a world dominated by relentless competition." The scene shifts to a men's basketball game. "For all their down side, males are worth the trouble. Think of them as a female's insurance policy against losing her children to rapidly evolving threats like measles and the flu."

A viewer who knows nothing else about this topic would probably conclude at this point that scientists now know the answer to the question, "Why sex?" But the viewer would be mistaken—misled by an extraordinarily shallow and lopsided account of one of the most controversial topics in evolutionary biology.

The very existence of sexual reproduction presents a problem for Darwin's theory. The easiest way for an organism to reproduce is simply to divide asexually—to make a copy of itself. Bacteria are very successful at this. An organism that reproduces sexually, however, must divert precious energy into making sperm or egg cells; in the process, gene combinations that were quite useful beforehand are sometimes destroyed through "recombination." Then the organism must find a member of the opposite sex and mate with it successfully. From an evolutionary perspective, sex incurs considerable costs that must be offset by advantages to the organism. But what are those advantages?

Various theories have been proposed—including the genetic variability theory to which Vrijenhoek subscribes. But situations such as the one he studied are relatively rare; so why is sexual reproduction so widespread?

A comprehensive review of this topic in 1988 concluded: "While there might well be agreement about the importance of the problem of the evolution of sex, there is no consensus about where its solution lies." Ten years later, in September 1998, the journal *Science* devoted a special issue to the evolution of sex, and said essentially the same thing: Biologists "haven't solved the mystery of sex yet," partly because of "extremely lousy experimental data." And this was *after* Vrijenhoek had done his research on Mexican minnows. So evolutionary biologists continue to "scrounge for data to support one or the other of the warring theories of sex." According to *Science:* "How sex began and why it thrived remain a mystery."[1]

For people who think biology is boring, the lively debate over the biological reasons for sex could make the field much more interesting. But people won't learn anything about this controversy from *Evolution*.

C. The Origin of Sex and Gender Differences

The narrator continues: "If the reason for sex is a bit less mysterious these days, its origins remain much more speculative." A whimsical cartoon animation fills the screen. "Some believe it all got started billions of years ago, with two single-celled creatures sharing a chance encounter in the primordial night. They meet, and genes are exchanged. That's what sex is all about. The moment is brief, but it leaves them a little bit stronger, a little more likely to survive and reproduce. Males and females came later, when random change produced a creature that was small and fast, which turned out to be an evolutionary advantage. Organisms with reproductive cells like that are called males. Their goal is to find organisms with a different specialty—providing the nutrients life requires. They're called females. These early pioneers evolved in time into sperm and eggs."

The difference between sperm and eggs is then extrapolated into an account of the differences between male and female sexual attitudes. "Males produce sperm by the millions, with so many potential offspring it doesn't pay to be fussy about eggs. A better strategy is to try to fertilize every egg you can. Eggs are more complex than sperm and take a larger investment of energy. Females make only a limited number of them. Fewer eggs mean fewer chances to pass on genes, and that means females—unlike males—do better if they're choosy. At a deep biological level, males and females want different things, regardless of how things appear on the surface."

The camera pans over a man and woman—apparently naked beneath a sheet that is pulled back to expose as much flesh as network policy permits—who are engaging in sexual foreplay. "Small sperm versus large eggs," the narrator continues. "Quantity versus quality. These are the evolutionary roots of the war between the sexes." This war "can explain a lot about how species evolve, and why they look and act the way they do. Charles Darwin was the first to recognize the evolutionary significance of sex." These statements are graphically illustrated with scenes of various animals engaging in sexual intercourse.

According to the narrator, Darwin's theory of natural selection could explain why "any trait that improves an individual's chances of survival should spread through the entire population. But it offered no help in explaining the wild extravagances found throughout nature, like the peacock's tail."

"It is hard to see a peacock's tail as something other than an impediment to his survival," the narrator says. "Theologians of [Darwin's] day argued that God created ornate flowers and feathers to inspire man's wondering devotion. Darwin was convinced there had to be an evolutionary explanation—just as there had to be an evolutionary explanation for why so many of nature's ornaments are seen only on males."

To solve the problem, Darwin formulated his theory of sexual selection, which psychologist Geoffrey F. Miller calls "Darwin's most ingenious idea." According to Miller, "these ornaments are not for our good. They're to advertise each individual's fitness, its goodness as a mate, to the opposite sex."

We watch scenes of male animals fighting, and a movie scene of John Travolta putting on seductive clothing, as the narrator tells us that females engage in choice, while males engage in competition. Male competition may take the form of fighting, we are told, or it may take "the path of the peacock—seduction through sexual display. This is where female choice comes in."

Commentators take turns emphasizing that the idea of female choice was controversial in Victorian England. Only a century later was this aspect of Darwin's theory tested in peacocks, when experiments showed that males with bigger, flashier tails tend to attract more mates and to have longer-surviving offspring. "It's all a logical consequence," the narrator says, "of the differing reproductive strategies of males, who have lots of sperm, and females, who have fewer eggs."

As the narrator acknowledges, however, all of this is speculation. Science is supposed to rest on evidence, but a whimsical cartoon animation about the evolutionary origin of sex is not evidence. In fact, most of what we have just seen is what evolutionary biologist Stephen Jay Gould would call a "just-so story." About a hundred years ago, Rudyard Kipling wrote a children's book by that name which recounted entertaining but scientifically meaningless stories about how leopards got their spots, and other things. In just-so stories, according to Gould, "virtuosity in invention replaces testability as the criterion for acceptance." *Evolution* is telling us just-so stories, yet we are expected to regard them as scientific and to draw two far-reaching conclusions: First, that sex originated in "random change;" and second, that men and women behave as they do because the former have lots of small, fast sperm and the latter have a few large, complex eggs.[2]

Not only does *Evolution* rely on just-so stories, but it also frames yet another scientific issue in religious terms. We have just been presented with two competing ideas: the idea that "God created ornate flowers and feathers to inspire man's wondering devotion," and the idea that peacocks' tails are a result of Darwinian sexual selection. Then we are told that the tendency of male peacocks with big, ornate tails to attract more mates and have longer-surviving offspring is "a logical consequence of the differing reproductive strategies" of males and females—as though that somehow refutes the idea that God created ornate flowers and feathers. Why not just present the evidence, and leave religion out of it? Once again, *Evolution* goes out of its way to speak to the religious realm.

C. Marriage and Family

"But the goal isn't just to have offspring," the narrator continues. "The young have to survive long enough to have their own offspring. Sometimes that requires paying as much attention to behavioral traits as to physical ones." In a scene from an old movie, Katherine Hepburn has difficulty choosing between sexiness and dependability in her man. This "mirrors a deep biological dilemma," the narrator tells us. "For some species, the chances of offspring surviving increase if a female chooses a mate who'll stick around over the one with the best genes."

We watch songbirds in which males and females share the job of parenting. The narrator tells us that the female needs the male's help, but the male will stay home only if he believes the chicks he's helping to raise are his own. "The result is monogamy—a social solution to a biological dilemma." The scene shifts to the human family we met earlier. The husband and wife affirm their commitment to their adopted child and to each other as the narrator says that "monogamy isn't easy to maintain. While some evolutionary forces encourage it, others threaten the family values that are at its core."

"Songbirds are unusually monogamous," the narrator continues. "But even as they pair off and set up nests, inevitably some of them are lusting after their neighbors." Cornell University behavioral ecologist Stephen T. Emlen explains how a female songbird returning from migration sometimes has to settle for "a fairly low-quality male" in comparison with her neighbors. The female is now torn, according to Emlen, between a desire to have a faithful mate who will help her raise her young, and a desire to have her chicks sired by a male of higher genetic quality. "Cheating, at least for certain female songbirds," says the narrator, "gives their chicks better genes, and therefore a better chance of surviving until they can reproduce."

For a species of bird in Panama, the narrator tells us, "survival of chicks is so uncertain it's led to an amazing gender role reversal." So many chicks are lost to crocodiles that females leave their eggs for the males to raise, and go off to reproduce again. "Now it's the females who care more about quantity than quality. Now it's the females who fight over mates. Over time, they've taken on traditionally male characteristics." Emlen explains that the females of this species are aggressively territorial, and try to attract "harems" of four or five males.

The narrator continues: "So here is an evolutionary revelation about gender: Male and female roles are not set in stone. They're largely determined by which sex competes for mates, and which invests in the young."

Wait a minute! Just a few scenes earlier, we were told that males and females behave as they do because the former have lots of small, fast sperm and the latter have a few large, complex eggs. Now we are told that the behavior of males and females depends on which sex competes for mates and which invests in the young. Yet the males of these Panamanian birds still produce sperm, and the females still lay eggs. Apparently, the reasons for male and female behavior are not as simple as *Evolution* would like us to think.

D. Chimpanzees and Bonobos

"Solving the problem of passing on genes can even trigger the emergence of new species," the narrator says. "Sometimes what separates species is more social than physical, as it is with our closest relatives, chimpanzees and bonobos." Chimpanzees and bonobos look alike, live in similar environments, and eat similar food. Chimpanzees are very pugnacious, however, while bonobos are essentially peaceful. "Bonobos are predisposed to make love, not war," the narrator tells us, while two of them copulate on screen.

We watch wild chimpanzees fighting—and occasionally stopping just long enough to have sex. Then we visit the San Diego Wild Animal Park, where we witness bonobos having heterosexual and homosexual intercourse "in every way imaginable"—with a running commentary to make sure we don't miss anything. No doubt many of the teenagers who watch the *Evolution* series in their public school science classrooms will be hugely entertained by these promiscuous primates.

So why do chimpanzees make war, and occasionally love, while bonobos seem to make only love? The answer, we are told, is female solidarity: Bonobo females are "able to form alliances with each other and cooperatively dominate males. And this changes the whole balance of power and the whole social dynamic in the group, and makes it radically different from chimpanzees."

Why have bonobos evolved this strategy, and chimpanzees haven't? "It looks as though a relatively simple change in the feeding ecology is responsible for this dramatic difference in sexual behavior," we are told. Bonobos live in forests where they can forage for food on the ground. Although there are chimpanzees that live in similar forests, those forests are also occupied by gorillas. The gorillas eat the food on the ground, leaving the chimpanzees dependent on fruit trees. So female chimpanzees have to forage intermittently and alone, without the opportunities for social interactions enjoyed by female bonobos.

The narrator concludes: "The simple fact that there was food available on the ground appears to have been the force that drove the evolution of bonobos." He speculates that some chimpanzees evolved into bonobos about two million years ago because they were able to forage on the ground after a drought killed the gorillas. We are told that if our own ancestors had experienced the same conditions that supposedly led to the evolution of bonobos, "we might have evolved to be a totally different, more peaceful, less violent, more sexual species."

But does "more sexual" really mean "less violent"? Violent chimpanzees seem to be just as sexual as peaceful bonobos. The slogan "make love, not war" sounds good, but sex and love are not synonymous—a lesson which many people have learned only after deep personal tragedies. What is *Evolution* trying to teach students here?

In any case, this account of the evolutionary origin of bonobos is another just-so story. We have two species with differing behavior patterns. But did differences in behavior lead to the origin of two species? Or did the two species originate in some other manner, with different behavior patterns from the start? Did opportunities for ground foraging produce female solidarity, which in turn established social peace? Or was it the other way around? How can we know? Where's the evidence?

As we watch actors in hairy costumes cross an African plain, the narrator acknowledges that this theory of the origin of bonobos "is little more than interesting speculation. But the idea behind it is consistent with a growing but controversial body of scientific thought that claims much of present-day human behavior is rooted in our distant past." That controversial body of thought is "evolutionary psychology," and it claims that modern human behavior patterns were formed under primitive conditions on the plains of Africa millions of years ago.

E. Evolutionary Psychology

"Evolutionary psychologists begin by pointing out," says the narrator, "that regardless of the culture in which we grow up, we all tend to respond the same way to a surprising variety of things. Most of us find spiders unpleasant, certain body types sexy, and particular smells disgusting. All, they say, are legacies of our evolutionary past."

Researchers conduct an experiment in which young men sleep in the same T-shirts night after night, and put the shirts in plastic bags during the day. Then a panel of young women smells the shirts and rates their sex appeal. According to the researchers, the women consistently prefer the shirts from those men who differ most from them in their immune genes. "From an evolutionary perspective, this makes sense," the narrator tells us. "Choosing a mate with different immune genes gives offspring a greater protection against viruses, parasites, and other pathogens."

Facial beauty, we are told, is simply a collection of subconscious biological cues that let us know whether a potential mate is genetically desirable. In one experiment women choose attractive male faces from a computer program, and they tend to prefer more masculine faces at those times of the month when they would most likely become pregnant.

Experiments such as these are conducted by evolutionary psychologists who think human behavior must be explained within a Darwinian framework. But of all the claims by evolutionary psychologists, the narrator tells us, none are more sweeping than those made by Geoffrey Miller: "He believes the human brain, like the peacock's magnificent tail, is an extravagance that evolved—at least in part—to help us attract a mate, and pass on genes."

"The human brain," Miller says, "is the most complex system in the known universe. It's wildly in excess of what it seems like we would need to survive on the plains of Africa. In fact, the human brain seems so excessive that a lot of people who believe in evolution applied to plants and animals have real trouble imagining how natural selection produced the human brain."

We watch ants scurrying along a log as Miller continues: "All the other species on the planet seem to get by with relatively small, simple nervous systems that seem tightly optimized just to do what the species needs to do to get by." The scene shifts to a yet another hunched-over actor in a hairy costume. "I think people are perfectly sensible in being skeptical about the ability of selection for survival to account for the human brain. I think there was a sort of guidance happening, there was a sort of decision-making process that was selecting our brains. But it wasn't God, it was our ancestors. They were choosing their sexual partners for their brains, for their behavior, during courtship." We see an apeman-like figure squatting on a narrow ledge. "And I think our semi-intelligent ancestors were the guiding force, they were the guiding hand, in human evolution."

We return to the husband and wife with their adopted child. "When choosing a mate, we still notice beauty," the narrator says, "but what really counts is how someone thinks, feels and acts. All of these are products of the brain." After watching an old film clip of long-nosed Cyrano de Bergerac professing his love to Roxanne, we are told: "It's brains, not beauty, that win her heart."

Miller continues: "There are all sorts of things that mess up brains. And paradoxically, for that reason, brains make really good indicators of how fit you are during courtship. In fact, they're probably better indicators of that even than, than a peacock's tail is about how fit a peacock is."

But Darwin formulated his theory of sexual selection to explain the striking *differences* we see between the males and females of some species. Sexual selection is supposed to explain why male peacocks have large, colorful tails—and females don't. But men's brains are not significantly larger or more colorful than women's brains. Miller is quite right when he says it is implausible to attribute the human brain to natural selection, because our brain is so much more than what creatures would have needed to survive on the plains of prehistoric Africa. But attributing the human brain to sexual selection is even more implausible

Nevertheless, we are told that Miller's hypothesis is "an intriguing idea," because "it's not the same old saw of tool use, language, culture—it's something entirely different." This is consistent with Stephen Jay Gould's claim, quoted above, that "virtuosity in invention" is "the criterion for acceptance" among evolutionary psychologists. *Evolution* is telling us just-so stories.

The scene shifts to a performance of the "Hallelujah Chorus" from Handel's *Messiah*, as the narrator says: "Miller is just getting started when he argues that the size of our brains can be attributed to our ancestors' sexual choices. He's also convinced that artistic expression, no matter how sublime, has its roots in our desire to impress the opposite sex. And that includes music, art, the poetic and storytelling uses of language—even a good sense of humor. According to Miller, they all stem from our instincts for sexual display."

"I think," says Miller, "when a lot of people produce cultural displays, what they're doing in a sense is exercising these, these sexual instincts for impressing the opposite sex. They're not doing it consciously, but what they're doing is investing their products with an awful lot of information about themselves." We watch part of a ballet, and Miller concludes: "I think the capacity for artistic creativity is there because our ancestors valued it when they were making their sexual choices."

So Miller sees all of human culture as a by-product of sexual urges—just as Freudian psychology did. But Freudian psychology is no longer considered good science. "Freud's views lost credibility," wrote University of Chicago evolutionary biologist Jerry A. Coyne in 2000, "when people realized that they were not at all based on science, but were really an ideological edifice, a myth about human life, that was utterly resistant to scientific refutation. By judicious manipulation, every possible observation of human behavior could be (and was) fitted into the Freudian framework. The same trick is now being perpetrated by the evolutionary psychologists. They, too, deal in their own dogmas, and not in propositions of science."

Coyne was criticizing evolutionary psychology in general. But many biologists also criticize Miller's specific ideas about the evolution of the human brain. "How does one actually test these ideas?" wrote University of Sheffield behavioral ecologist Tim Birkhead in a 2000 review of Miller's work. "Without a concerted effort to do this, evolutionary psychology will remain in the realms of armchair entertainment rather than real science." In another review of Miller's work, American Museum of Natural History paleoanthropologist Ian Tattersall wrote: "In the end we are looking here at a product of the storyteller's art, not of science."[3]

Although this episode acknowledges that Miller's ideas are controversial, it presents them uncritically, without mentioning the fact that many biologists don't even consider them scientific. Like the controversy over the biological reasons for sexual reproduction, the controversy over the scientific status of evolutionary psychology is completely ignored by *Evolution*.

This shallow and lopsided account of evolutionary psychology also has a religious—or rather anti-religious—component. Despite the lack of evidence for his hypothesis, Miller says he is confident that the reason we have our brains "wasn't God, it was our ancestors. They were choosing their sexual partners for their brains, for their behavior, during courtship." And to illustrate Miller's claim that artistic creativity is reducible to our ancestors' sexual choices, *Evolution* chooses—of all things!—the "Hallelujah Chorus."

God and the Messiah. More religion. Despite the assurance by *Evolution*'s producers that they would avoid "the religious realm," they can't seem to stay away from it.

F. Into the Future?

As we watch a collage of scenes from throughout the episode, the narrator reminds us: "Sex is at the heart of evolution. The process of mixing and passing on genes produces variation, that helps species meet the challenge of life in a competitive world. Sexually selected variations are those that help individuals find mates, and successfully raise young. That's how, for humans, sex became fun, and parenting rewarding."

The scene shifts once more to actors wearing apeman costumes, and the narrator continues: "Those of our ancestors who took pleasure from sex, and satisfaction from parenting, had more surviving offspring than those who didn't. That was true generation after generation. These traits are now almost universal. Even if we choose not to have children, we still enjoy sex. And even when we adopt a child who doesn't carry our genes, we can still find parenting rewarding."

We look in once again on the husband and wife who adopted a baby. Humans, we are told, are the only species that will care for biologically unrelated children over the long term. "Humans are unique," the narrator says. "We are a product of evolution. But we've taken the first tentative steps towards controlling our evolutionary destiny. It's a brave new world we're entering. Only time will tell if we'll be as successful at guiding our future as evolution has been."

So parenting is a good thing—even when it is no longer connected to sex, when it no longer serves the evolutionary purpose of passing on genes. This episode leaves us with some strangely mixed messages. The reproductive behavior of males and females is due to the differences between sperm and eggs—except in certain Panamanian birds, when it isn't. Sexual selection produces striking differences between males and females—except in the evolution of the human brain, when it doesn't. And parenting must be understood in an evolutionary perspective—except in human families with adopted children, when it mustn't.

By ending with a husband and wife who are raising a child unrelated to them, this episode actually raises an issue that is even more of a problem for Darwin's theory than the existence of sex: altruism. Altruism is defined biologically as increasing the fitness of another at the expense of one's own fitness. But an altruist thereby reduces his or her own chances for survival; so in the context of evolutionary theory, altruism should not survive or evolve. Yet altruistic people exist—as we have seen in this episode. Harvard University sociobiologist Edward O. Wilson (founder of the discipline which gave rise to evolutionary psychology) has called this "the culminating mystery of all biology." Why doesn't *Evolution* even mention it?[4]

So this episode completely ignores three major controversies raging beneath the surface of its topic: one over the biological reasons for sex, another over the scientific status of evolutionary psychology, and still another over the mystery of altruism. Rather than educate viewers about what's really going on in biology, *Evolution* emphasizes two simple-minded messages: God is out, and sex is in. Sex is more important than life itself. Sex is our immortality. Sex is why we have big brains. Instead of providing us with solid scientific evidence, or an honest treatment of serious scientific controversies, this episode relies on just-so stories and sex scenes.

Why Sex? It's a fascinating question. But viewers of *Evolution* may find themselves asking: Why so much sex—and so little science?

Notes

1. For a review of the research on Mexican minnows, see Robert C. Vrijenhoek, "Animal Clones and Diversity," *BioScience* (August, 1998), at:

    http://www.findarticles.com/cf_0/m1042/n8_v48/21007759/p1/article.jhtml

    The 1988 study that reported no consensus on solving the problem of sex also reported: "A survey of evolutionary biologists would doubtless come up with a consensus that the elucidation of the selective pressures responsible for the origin and maintenance of sex is a 'big' (maybe the 'biggest') unsolved problem in evolutionary biology." Richard E. Michod and Bruce R. Levin, *The Evolution of Sex: An Examination of Current Ideas* (Sunderland, MA: Sinauer Associates, 1988), vii.

    The quotations from the 1998 special issue of *Science* are from Bernice Wuethrich, "Why Sex? Putting Theory to the Test," *Science* 281 (1998), 1980-1982. The same issue included the following articles of interest: Pamela Hines & Elizabeth Culotta, "The Evolution of Sex," *Science* 281 (1998), 1979; N. H. Barton & B. Charlesworth, "Why Sex and Recombination?" *Science* 281 (1998), 1986-1990.

    For more about the controversy, see: Lynn Margulis and Dorion Sagan, *What Is Sex?* (New York: Simon & Schuster, 1997)—see especially 121: "The Red Queen idea [which forms the basis of Vrijenhoek's hypothesis] is simply a cute name for a zoological myth"; and John Cartwright, *Evolution and Human Behavior* (London: Macmillan, 2000), especially 90-101.

2. The Stephen Jay Gould quotation is from "Sociobiology: the art of storytelling," *New Scientist* (November 16, 1978), 530.

3. The Coyne quotation is from Jerry A. Coyne, "Of Vice and Men: The fairy tales of evolutionary psychology," a review of Randy Thornhill and Craig Palmer's *A Natural History of Rape,* in *The New Republic* (April 3, 2000), last page. The entire review is available at:

    http://www.thenewrepublic.com/040300/coyne040300.html

    For a recent critique of sociobiology and evolutionary psychology for non-specialists, see Tom Bethell, "Against Sociobiology," *First Things* (January, 2001)

    http://print.firstthings.com/ftissues/ft0101/articles/bethell.html

    The Birkhead quotation is from Tim Birkhead, "Strictly for the birds," a review of Geoffrey Miller's book, *The Mating Mind* in *New Scientist* (May 13, 2000), 48-49; the Tattersall quotation is from Ian Tattersall, "Whatever turns you on," a review of Geoffrey Miller's book, *The Mating Mind,* in *The New York Times Book Review* (June 11, 2000).

    For another scientist's critique of evolutionary psychology, see University of Leicester geneticist Gabriel Dover's *Dear Mr. Darwin* (Berkeley: University of California Press, 2000), especially 44-45. Dover wrote:

This problem with just-so story telling is not some minor irritation.... The problem runs much deeper and wider, embracing many new disciplines of evolutionary psychology, Darwinian medicine, linguistics, biological ethics and sociobiology. Here quite vulgar explanations are offered, based on the crudest applications of selection theory, of why we humans are the way we are.... There seems to be no aspect of our psychological make-up that does not receive its supposed evolutionary explanation from the sorts of things our selfish genes forced us to do 200,000 to 500,000 years ago.... Not only is there the embarrassing spectacle of psychologists, philosophers and linguists rushing down the road of selfish genetic determinism, but we are also shackled with their self-imposed justification in giving 'scientific' respectability to complex behavioral phenomena in humans which we simply do not so far have the scientific tools and methodologies to investigate.

4. The E.O. Wilson quotation is from *Sociobiology* (Cambridge, MA: Harvard University Press, 1975), 382. The definition of altruism is taken from the abridged paperback edition (1980), 55.

For more about the challenge that altruism poses for evolutionary theory, see H. R. Holcomb, *Sociobiology, Sex, and Science* (Albany, NY: State University of New York Press, 1993); K. R. Monroe, *The Heart Of Altruism: Perceptions Of A Common Humanity* (Princeton: Princeton University Press, 1996); and H. Plotkin, *Evolution in Mind: An Introduction to Evolutionary Psychology* (Cambridge, MA: Harvard University Press, 1997).

EPISODE

6

The Mind's Big Bang

The emergence of art, technology, and society about 50,000 years ago. Hominid evolution and Neanderthals. Early human migrations. Language. Memes and how they now counteract biological evolution.

A. Cave Paintings, Stone Tools, and Fossil Skulls

An archaeologist crawling through a cave in France is "searching for a special moment in evolution," the narrator tells us, "an era cloaked in mystery, when with hardly a change in appearance, humans began behaving in ways they had never behaved before. He wants to find out how it was our ancestors became truly human." Where once there were just bare cave walls, suddenly there was art, technology, communication and culture. "The question is, What happened to make all this possible? How could it be, a species opened its eyes and burst into a new realm? How was it, human ancestors evolved a whole new way of seeing themselves? And because of this, in time transformed the planet?"

We fly over a mist-shrouded landscape. "The Great Rift Valley of East Africa," the narrator continues, "here is where the human story began. For millions of years, Africa was the landscape of human evolution. Across this terrain, an ancestral people survived, reproduced, and passed on who they were to succeeding generations. Without Africa, humanity as we know it might never have become."

We stop at a spot in the Great Rift Valley which was "once inhabited by hominids, before they were truly human." Smithsonian Institution paleoanthropologist Rick Potts clambers up a hillside looking for fossils and stone tools. "Now it's a site scientists visit to understand how people lived and what they thought about a million years ago." Potts digs up a stone axe, and a computer animation shows us primitive humans making such tools and using them to butcher a large animal.

"Here, across this terrain," the narrator says, "these paleolithic or ancient stone-tool people made one simple implement for nearly a million years." According to Potts, the stone axe he found was the Swiss Army knife of the paleolithic period. The people who made them, he says, made the same thing over and over, but they probably didn't speak to each other as we do. "They didn't have something that we have—the creativity, the innovation, the diversity of cultures that of course characterizes our own species."

Another computer animation takes us on a long journey through time. "On the tree of life," the narrator tells us, "human evolution began around six million years ago when hominids split off from the common ancestor they shared with chimpanzees. They descended from the trees about four million years ago and entered a new world. Two and a half million years ago with a modified hand they fashioned stone tools and began to depend more and more on a diet of meat." Thanks to the computer, this all happens before our very eyes.

"The size of their brains increased substantially," continues the narrator. "At about two million years ago they began to leave Africa. These early humans were successful for a while but in the end every one of them would become extinct. It wasn't until fifty or sixty thousand years ago that the first truly modern humans, our ancestors, left Africa." Human actors take the place of computer-generated figures. "They were hunter-gatherers, foraging for food living in small groups roaming the wide landscape, but they were different from their predecessors. They had begun to live a revolutionary new way of life."

A series of reconstructed skulls appears on the screen, starting with one that is very ape-like and ending with one from a modern human. "This lifestyle had been achieved over millions of years," says the narrator, "through the multiple processes of evolution—adaptation, competition, mutation, selection, and failure, punctuated by the occasional success. Ours was a routine story of evolution, of change over time, no different from the stories of so many other species, but it produced behavior new to the planet."

B. The Beginnings of Art and Technology

"Behavior changed very radically around fifty thousand years ago," we are told. A fossil skull appears, but "this hundred-thousand-year-old human did not behave like us." Fully modern human behavior, we are told, included the making of a wide range of artifacts, such as art and jewelry.

According to Massachusetts Institute of Technology psychologist Steven Pinker: "In a sense, we're all Africans." He explains that human babies from all over the world have the same basic ability to learn languages, how to count, and how to make and use tools. "It suggests," says Pinker, "that the distinctively human parts of our intelligence were in place before our ancestors split off into the different continents."

"After leaving Africa some fifty or sixty thousand years ago," the narrator continues, "this fully modern species headed east, into Asia, and even to Australia. Others followed the coast of the Mediterranean north, dispersing into the hills and leaving behind evidence that their minds were unique to this planet." A small boat rounds a point of land, and we are introduced to University of Arizona anthropologists Mary C. Stiner and Steven L. Kuhn. Stiner and Kuhn are "excavating a home that these early immigrants occupied," the Ucagizli Cave in Turkey, "one of the earliest modern human living sites." To their surprise they find an abundance of ornaments, including beads dated at forty-three thousand years ago—making them "the oldest beads found so far anywhere in the world."

Beads, however, are of no practical use in a hunter-gatherer society. "They would suggest," says the narrator, "that those who lived in this formidable place had more on their minds than straightforward survival. What could have been so important about these beads, and what can they tell us about these early days of modern humans?" By the time humans had migrated to what is now southern France, we are told, they were mass-producing beads with a distinctive grinding technique. "Beads are an artifact of the mind's big bang," the narrator says. "They are evidence of our creative and cultural beginnings. They suggest a time when humans began relating their own social groups to groups of other humans."

By wearing ornaments such as beads, we are told, these ancient humans were "expressing social relationships." And that was "very new in human evolution." Some prehistoric jewelry is displayed. "Humans using technology in the service of social identity," the narrator says. "This was momentous." A shadowy figure pulls a burning stick out of a fire. "This transformation of our minds began in Africa, and it left a trail of evidence as far away as Australia. But the clues are most abundant in Europe."

The scene shifts to a green countryside. In Europe, the narrator tells us, "humans encountered another species of hominid—a species almost identical to us—but not quite." A drawing shows massive, hairy, almost-human figures. "It's this 'not quite' that tells us about our selective advantage. We call these ancient Europeans Neanderthals."

Neanderthals were bigger than we are, and they had receding chins and fore-heads. Most notably, Neanderthal burial sites were simple compared with ours, and these creatures apparently did not use pictures or symbols. "In contrast," the narrator says, "modern humans seemed to be treating their dead with extreme care."

A researcher compares the heavy stone-tipped spears used by Neanderthals with the lighter antler-tipped spears used by modern humans, and he concludes that modern humans were smarter and more technologically advanced. He says that Neanderthal culture was relatively unchanging, but modern human culture changed rapidly after it first appeared about fifty thousand years ago. This sug-gests that modern humans—unlike Neanderthals—were able to improve on what went before, from one generation to the next, and the narrator calls this ability "a strategic advantage."

"Improved technology suggests much," says the narrator, "especially humans' emerging ability to transmit information over great distances and through the realms of time." We see more drawings of massive, hairy figures, as we are told that Neanderthals lived in small, isolated groups. "For modern humans," the narrator continues, "portable art may have served as a means of communication—some of it travelling many hundreds of miles from where it had been created." Thus modern humans, unlike Neanderthals, were able to establish a far-ranging culture.

So we are told that we have evidence of modern humans—people, like us—starting about fifty thousand years ago. But a few minutes ago we were told that "for millions of years . . .an ancestral people survived, reproduced, and passed on who they were." We were told that scientists visit the East African rift valley to "understand how people lived and what they thought about a million years ago." Clearly, the "people" who lived a million or more years ago were not people in the ordinary sense of the word. Among other things, they didn't have language, technology or art. Apparently, not even Neanderthals were people in the ordinary sense of the word. Why, then, does *Evolution* call million-year-old animals "people"?

It seems we are being conditioned to accept the Darwinian account of human origins before we even see the evidence. The truth is that fossil skulls are reconstructed from fragmentary evidence, sometimes collected from different sites. The reconstruction and interpretation of such skulls is so controversial that paleoanthropologists—people who study human origins—call them "bones of contention." And even if individual specimens were not so controversial, it would still be impossible to arrange them confidently in a series of ancestors and descendants. As we saw in Episode Two, the picture of human evolution we are being shown here carries the same validity as a bedtime story. According to evolutionist Henry Gee, chief science writer for *Nature*, it is "a completely human invention created after the fact, shaped to accord with human prejudices."[1]

Since the evidence for human evolution is so weak, *Evolution* simply expects us to take the Darwinian account for granted. Instead of acknowledging that we have no way of knowing whether these extinct creatures were related to us—much less "what they thought"—*Evolution* calls them "people" and tells us more just-so stories.

We find ourselves in a cave, where an archaeologist shows us a "spit-painting" technique that may have been used by ancient humans to produce the cave art we find today. He speculates about why people made cave-paintings; he also finds evidence that cave-dwellers may have made music. As we leave the cave, the narrator says: "So below and above ground, our ancestors were refining technology and art, and communicating in complex ways. And it appears as if these changes occurred almost overnight. How could it have happened?"

C. Brain Mutations and Child Development

Reflections of primitive-looking figures shimmer on some water, as a man flanked by fossil skulls says: "My own view is that there was a brain change—that there was a genetic change that promoted the fully modern human brain, that allowed the kind of innovation and invention—the ability to innovate and invent—that is a characteristic of modern humans. If you accept the idea that there was a neurological change fifty thousand years ago, and that this was rooted in biology, it would just become the latest and most recent in a long series of mutations on which natural selection operated to produce the human species as we understand it today."

Steven Pinker returns to tell us: "It's very likely that the changes in the brain didn't happen overnight. There wasn't one magical mutation that miraculously allowed us to speak and to walk upright and to cooperate with one another and to figure out how the world works." Primitive-looking figures walk down a stream-bed as Pinker continues: "Evolution doesn't work that way. It would be stagger-ingly improbable for one mutation to do all of that. Chances are, there were lots and lots of mutations over a span of tens, maybe even hundreds of thousands of years that fine-tuned and sculpted the brain to give it all the magnificent powers that it has today."

But Pinker knows nothing about "mutations" that could have "fine-tuned and sculpted the brain." Nobody does. There is no evidence that genetic mutations can do such a thing—not even over thousands of years. In fact, scientists have only vague ideas of how genes affect brain development, even in simple animals. And all known mutations that affect development—such as the *Antennapedia* mutation described in Episode Two—are harmful. As Cambridge University geneticist David L. Stern wrote in 2000, "one of the oldest problems in evolutionary biol-ogy"—the generation of relevant variations by mutations—"remains largely unsolved." Pinker's statement that mutations could have given the brain "all the magnificent powers that it has today" is sheer speculation.[2]

We watch a cartoon animation of the brain, and Pinker tells us that "a lot of our evolution consisted not just in getting more of this stuff, but in wiring it in precise ways to support intelligence." The narrator adds: "So it may not have been the size of the human brain, but its wiring, that endowed us with substantial new skills."

Chimpanzees scamper through the trees in a Ugandan forest, as the narrator says that one of those new skills might have been "the knack for living a complex social life. Here in East Africa, chimpanzees show us how we might have inter-acted with others *before* the mind's big bang." Chimpanzees can induce other chimpanzees to behave socially only through direct physical force. "But after six million years of separate evolution," the narrator says, "humans have acquired a significant advantage"—language.

We go to the University of St. Andrews in Scotland, where psychologist Andrew Whiten studies learning patterns in young children. According to the narrator, Whiten found that through the age of three "a child cannot ascribe actions, motives and beliefs to others. But by the age of five, the child's brain has developed the capacity for stepping into someone else's mind." Chimpanzees, however, never reach this stage.

Actors portraying primitive people stand over a blazing fire, as Whiten says: "In a society of humans, being socially competent really counts. Being socially competent allows you ultimately to out-compete others, to gain better access to resources, the best mates. And in those kind of societies, it seems the brain can be more important than brawn. So it's potentially a very powerful evolutionary force, because it's driving a kind of upward spiral. Social complexity begets greater social intelligence; social intelligence presents even greater problems to the individuals in the next generation; and they have to become more socially complex."

But the "evolution" Whiten describes here is social progress—not evolution in the Darwinian sense of new species originating from a common ancestor through natural selection. And as interesting as his research in developmental psychology may be, it tells us nothing about the *origin* of our ability to "ascribe actions, motives and beliefs" to others. Of course it is advantageous to understand how others think. But how did that ability originate? Nothing that has been presented here helps to answer that question.

D. Language

"Complex social relationships," the narrator adds. "A theory of mind. These are qualities we associate with modern humans. But how could we practice any of them without language?" The camera focuses on people's mouths as they speak. "With language we can relive the past, ponder the future, teach our children, tell secrets, manipulate crowds. But imagine a world without language."

We travel to Managua, Nicaragua, where we meet "Mary No-Name"—a woman who has been deaf since birth. Years ago, U.S. experts went to Managua to teach standard sign-language to deaf children who had just been brought in from isolated villages, but they failed. It turned out that the children developed their own sign-language instead, without any help from the experts. Apparently, the only stimulus they had needed was to meet other deaf people with whom they wanted to communicate.

"Might this moment," the narrator asks, "resemble what happened around fifty thousand years ago—the turning-point that led to the explosion of human creativity?" A girl gestures expressively with her arms. "Language does not need a voice," the narrator says, "it is our legacy, an inevitability of being human. Today, we still don't know exactly when language evolved—when it opened the door to our phenomenal success as a species."

"While many species can communicate, even vocalize," the narrator continues, "only human languages are driven by complex rules." All human languages have these rules, called syntax, which enable us to organize information hierarchically, and to construct sentences.

According to Oxford University zoologist Richard Dawkins, those of our ancestors most gifted with the tools of language might have been those to prosper. We visit Dawkins as he waters his garden. "We don't know when language started," says Dawkins, "but as soon as language did start it provided an environment in which those individuals who were genetically best equipped to thrive and survive and succeed in an environment dominated by language were the ones who left the most offspring. And that probably—in our forefathers—that probably led to an improvement in the ability to use language."

Once again we find ourselves in the company of primitive-looking people, squatting next to a blazing fire. "What, exactly, was the evolutionary purpose of language?" the narrator asks. "Was it to discuss waterholes, weapons, and what lay over the hill? Or might it have had another advantage?"

In the passenger car of a train somewhere in the United Kingdom, University of Liverpool evolutionary biologist Robin Dunbar is eavesdropping on other people's conversations. "The kind of situations we're looking for to study language," he says, "[were] just the sort of natural spaces where you would have a conversation—a very informal, relaxed conversation—with friends." According to Dunbar, the standard view of people who study language is that its primary function is "the transmission of technically complex information. This is what I kind of call the Einstein and Shakespeare version of language." Dunbar finds, however, that "what people talk about on a day-to-day basis, back there in their homes, or on the street, or over the garden fence, then it's about social relationships."

The narrator says: "Two-thirds of all our conversations, Robin Dunbar believes, are dedicated to gossip. Throughout human evolution, could nature have selected not just for the fittest, but for those with the most refined social skills?" Dunbar continues: "What language does—the bottom line, if you like—is it just allows us to hold big groups together. It's like kind of opening a window of opportunity. Suddenly there's all sorts of other things you can do with it. Because you can use it to solicit information about third parties so you can now see what happened when you weren't actually present at the time." This gives us an advantage over monkeys and apes, because "if they don't see it, they don't know about it, and they never will."

Steven Pinker, however, thinks Dunbar's view is only part of the story: "Gossip is certainly one of the things that language is useful for, because it's always handy to know who needs a favor, who can offer a favor, who's available, who's under the protection of a jealous spouse." But "there are all kinds of ways that language can be useful. Gossip, I think, is just one of them."

So to be human, we are told, is to have language—even if it is language without speech. And language is useful in many ways—including holding together big groups, gossiping, and communicating technical information. But what do language and evolution have to do with each other?

Although languages certainly change over time, they do not evolve in the sense of becoming more complex. Languages cannot be ranked in order from primitive to more advanced. According to the *Cambridge Encyclopedia of Language:* "Anthropologically speaking, the human race can be said to have evolved from primitive to civilized states, but there is no sign of language having gone through the same kind of evolution." Complex language simply appeared, fully formed, with the first humans. There is no evidence for Dawkins's claim that a Darwinian process "led to an improvement in the ability to use language."

Furthermore, nobody knows how language originated. According to the *Cambridge Encyclopedia of Language:* "For centuries, people have speculated over the origins of human language. . . . [but] the quest is a fruitless one. Each generation asks the same questions, and reaches the same impasse—the absence of any evidence relating to the matter, given the vast, distant time-scale involved. We have no direct knowledge of the origins and early development of language, nor is it easy to imagine how such knowledge might ever be obtained. We can only speculate, arrive at our own conclusions, and remain dissatisfied."[3]

So language hasn't evolved in complexity since it originated, and its origin remains mysterious. We are told that language may have been the evolutionary turning-point that gave rise to modern humans. But in Episode Two we were told that "walking on two feet" may have been that turning-point, and in Episode Five we were told that our ancestors' "choosing their sexual partners for their brains" may have been that turning-point. The truth is that all of these are mere speculation. From an evolutionary perspective, the origin of language, the brain, and the human species itself remain mysterious. The question asked at the beginning of this episode—"What happened to make all this possible?"—remains unanswered.

This doesn't deter *Evolution*, however, from piling speculation on speculation: "Language," the narrator concludes, is "the force that made modern human culture possible, and that today tells us who we are, how we belong, where we're bound. Language, according to Richard Dawkins, is also central to a new and powerful evolutionary force."

E. Memes

"As far as a human lifetime is concerned," says Dawkins, "the only kind of evolutionary change we're likely to see very much of is not genetic evolution at all, it's cultural evolution. And if we put a Darwinian spin on that, then we're going to be talking about the differential survival of memes, as opposed to genes."

According to psychologist Susan Blackmore, of the University of the West of England: "Memes are ideas, habits, skills, gestures, stories, songs—anything which we pass from person to person by imitation. We copy them. Now, just as genes are copied inside all the cells of our body and passed on in reproduction, memes are copied by our brains and our behavior and they're passed from person to person. And I think what happens is, just as the competition between genes shapes all of biological evolution, so it's the competition between memes that shapes our minds and our cultures. So it's absolutely essential to understanding human nature that we take account of memes."

The narrator says that Blackmore "believes memes have been the forces driving human evolution, especially since the mind's big bang some fifty thousand years ago. She sees ideas, prejudices, trends and breakthroughs behaving much like genes—self-replicating and accumulating from mind to mind, society to society, generation to generation. Memes are the building-blocks of a new kind of evolution." Richard Dawkins adds: "If units of culture replicate themselves in something like the same way as DNA molecules replicate themselves, then we have the possibility of a completely new kind of Darwinism."

"Changes in the human lifestyle for the last fifty thousand years," says Steven Pinker, "have had very little to do with any biological change in our brains. The reason that we live differently today from the way the cavemen lived is not because we have better brains but because we've been accumulating all of the thousands of discoveries that our ancestors have made, and we have the benefit of a huge history of inventions that we communicate non-genetically, through language, through documents, through customs."

"Memes can be more than passing fads," the narrator tells us. "They can be titanic. They can modify the world, revolutionize life, even suppress the forces of biological evolution. Consider insulin, one such meme, now some eighty years old." We meet Jared, a fourteen-year-old diabetic, who says he would probably be dead without insulin. "Before the 1920s," the narrator continues, as we look at old photographs of sickly youngsters, "individuals like Jared would have died as children—never to reach the age of reproduction, never to pass on their genes. Now young diabetics are no longer condemned to death. Insulin, an idea that became a medicine, is just one more meme that helps modern humans elude the forces of evolution. It and so many other scientific breakthroughs provide us with new ways to survive."

Jared and his friends go on their way, and Pinker remarks: "A lot of the creations of the brain can actually make up for physical deficiencies and could actually change the course of evolution." To those who might say that it is unwise to interfere with evolution in this way, Pinker responds that for thousands of years humans have depended for survival on their own inventions, and this is simply "the way human evolution works." As we watch a collage of cultural diversity, the narrator adds: "Our rebellion against evolution has taken many forms. Call it culture, call it memes, call it memetic evolution—whatever. It makes every one of us this planet's best survivor—so far."

Blackmore concludes with: "Nowadays I would say that memetic evolution is going faster and faster, and it has almost entirely taken over from biological evolution. Not entirely, in a sense the two are going along hand in hand. For example, birth control—the memes of the pill and condoms and all these things have effects on the genes. In fact, they change quite dramatically, across the planet, which genes are getting passed on and which aren't. The more educated you are, the less children you have. That is memes fighting against genes. What's also going on now at the beginning of the twenty-first century is that the memes have suddenly made themselves a new home: the Internet." Blackmore says that although we thought we created the Internet, in fact we are its slaves. She regards this as an "inevitable" consequence of memetic evolution. "The memes are getting better and faster, and more and more, and creating as they go, better copying apparatus for their own copying. I don't know where that leaves us in the future."

So memes invented the Internet. And birth control. And insulin therapy for diabetics. And, it seems, everything else. What's left? Is there anything that is *not* a meme? A concept that describes everything describes nothing, and is too vague and too broad to be useful, especially in science. For just that reason, many scientists have criticized the meme concept. In Stephen Jay Gould's words, it is a "meaningless metaphor."

Instead of providing us with new insights, "meme" is apparently just a new label for familiar things like invention, science, art, and history. But inventors, scientists, artists, and historians did very well for centuries before Darwin and Dawkins came along. It was people, not memes, who discovered things like insulin, and invented things like the Internet, and created things like the Sistine Chapel, and shaped history by their actions. Blackmore's claim that memes "create" things is nonsense.

Dawkins talks about putting a "Darwinian spin" on "cultural evolution." But cultural evolution could just as well be called cultural history, as it was for centuries. Why put a Darwinian spin on it? According to Jerry Coyne (a Darwinist whose criticism of evolutionary psychology we met in Episode Five), memes are "but a flashy new wrapping around a parcel of old and conventional ideas." Could this simply be an attempt to force absolutely everything into a Darwinian framework? When Blackmore wrote a book promoting her ideas in 1999, Coyne called it "a work not of science, but of extreme advocacy."[4]

Whatever else we might say about memes, one thing is clear: They work *against* biological evolution. This "is not genetic evolution at all," says Dawkins. Memes are "our rebellion against evolution," says Pinker. It's "memes fighting against genes," says Blackmore. So if memes have any significance at all, they show that Darwin's theory of the origin of species is not as central as *Evolution* would like us to believe. Whether memetic evolution is unscientific, as its critics claim, or contrary to biological evolution, as its proponents claim, it is strangely out of place in this series.

Two people run across a plain and into the distance, as the narrator says: "For our species, as for all others, biological evolution has been the primary engine of change." The scene shifts to cave paintings. "But since the birth of culture some fifty thousand years ago, forces far more powerful have overtaken human evolution. The mind's big bang saw the birth of a new kind of change—not of the body, but of ideas. That means that for the future of humankind evolution may be no more than what we make of it."

The background music for these closing scenes is the hauntingly beautiful *kyrie eleison* from the *Missa Luba,* a Roman Catholic mass set to African music. Still more religion, in a series that promised to avoid "the religious realm."[5]

Kyrie eleison is Greek for "Lord, have mercy!"

Notes

1. On "bones of contention," see Roger Lewin, *Bones of Contention: Controversies in the Search for Human Origins,* Second Edition (Chicago: The University of Chicago Press, 1997). The Gee quotation is from Henry Gee, *In Search of Deep Time: Beyond the Fossil Record to a New History of Life* (New York: The Free Press, 1999), 32. For more on the controversial nature of paleoanthropology, see Jonathan Wells, *Icons of Evolution: Science or Myth?* (Washington, DC: Regnery Publishing, 2000), Chapter 11.

2. The Stern quotation is from David L. Stern, "Perspective: Evolutionary Developmental Biology and the Problem of Variation," *Evolution* 54 (2000), 1079.

    Pinker's glib statements about mutations ignore the fact that biologists know almost nothing about how genetic mutations might produce the sorts of changes evolution requires. In 1988, evolutionary biologists John Endler and Tracy McLellan wrote: "Although much is know about mutation, it is still largely a 'black box' relative to evolution." John A. Endler and Tracey McLellan, "The Processes of Evolution: Toward a Newer Synthesis," *Annual Review of Ecology and Systematics* 19 (1988), 397. Ten years later, evolutionary geneticist Allen Orr wrote: "Our understanding of the genetics of adaptation remains appallingly weak." H. Allen Orr, "The evolutionary genetics of adaptation: a simulation study," *Genetical Research* (Cambridge) 74 (1999), 212.

    A mutation that changes the direction of shell coiling in some snails is not known to be harmful, but it does not contribute anything to evolution. See M. H. Sturtevant, "Inheritance of direction of coiling in *Limnaea*," *Science* 58 (1932), 269-270.

3. The quotations from the *Cambridge Encyclopedia* are from David Crystal, *The Cambridge Encyclopedia of Language,* Second Edition. (Cambridge: Cambridge University Press, 1997), 6, 290. For a recent story on the controversy over linguistic evolution, see Bea Perks, "Linguists and evolutionists need to talk about linguistic

evolution," *The BioMedNet Magazine* (August 15, 2001), available at:

> http://news.bmn.com/news/story?day=010816&story=2

4. Richard Dawkins invented the term "meme" in the last chapter of *The Selfish Gene* (Oxford: Oxford University Press, 1976). Among other things, the idea of God is a meme. Dawkins wrote: "The survival value of the god meme in the meme pool results from its great psychological appeal. It provides a superficially plausible answer to deep and troubling questions about existence. . . God exists, if only in the form of a meme with high survival value, or infective power, in the environment provided by human culture." (207)

Stephen Jay Gould called memes a "meaningless metaphor" on a radio show November 11, 1996. See Susan Blackmore, "Memes, Minds and Selves," at:

> http://www.tribunes.com/tribune/art98/blac.htm#b

The Coyne quotations are from Jerry A. Coyne, "The self-centred meme," a review of Susan Blackmore's *The Meme Machine,* in *Nature* (April 29, 1999), 767-768.

For more criticism of the meme concept, see Mary Midgley, "Why Memes?" in Hilary Rose and Steven Rose (editors), *Alas, Poor Darwin: Arguments Against Evolutionary Psychology* (New York: Harmony Books, 2000), 79-99. See also:

> http://www.salon.com/july97/21st/meme970710.html

5. The *Missa Luba* is a mass originally sung in pure Congolese style by Les Troubadours du roi baudouin and directed by Father Guido Haazen. See:

> http://usrwww.mpx.com.au/~charles/Requiem/missa_luba.htm

The *kyrie eleison* in this episode probably comes from a newer version sung by the Mungano National Choir of Kenya and directed by Boniface Mganga; it is available as a Philips CD from Polygram Classics, New York, NY. About ten minutes earlier in the episode, just after Richard Dawkins speaks and some primitive-looking people are standing around a fire, the background music is the *agnus Dei* from the same *Missa Luba. Agnus Dei* is Latin for "the Lamb of God"—Jesus Christ.

7

What About God?

The creation-evolution controversy and U.S. science education. Biblical literalist Ken Ham. Students at Wheaton College struggle with their faith. A school board denies a petition to teach special creation alongside evolution.

A. The Creation-Evolution Controversy and U.S. Science Education

"The majesty of our Earth, the beauty of life," the narrator begins. "Are they the result of a natural process called evolution, or the work of a divine creator? This question is at the heart of a struggle that has threatened to tear our nation apart."

High school students file into a science classroom. A newspaper headline— "Collision in the classroom"—fills the screen. Answers in Genesis Executive Director Ken Ham gestures with a Bible. "For fundamentalist Christians like Ken Ham," the narrator continues, "evolution is an evil that must be fought." Ham says: "Oh, I think it's a war. It's a real battle between worldviews." We look in on a crowded school board hearing, and the narrator tells us: "For embattled teachers in Lafayette, Indiana, evolution is a truth that must be defended." One of those teachers says she doesn't think one side or the other will come out a victor. Then we join a round-table discussion among Christian students at Wheaton College in Illinois. According to the narrator, these students find evolution "an idea that is hard to accept." One student asks: "Where is God's place, if everything does have a natural cause?"

"For all of us," the narrator continues, "the future of religion, science, and science education are at stake in the creation-evolution debate. Today, even as science continues to provide evidence supporting the theory of evolution, for millions of Americans the most important question remains, What about God?"

Parents and children fill a church in Canton, Ohio, to hear Ken Ham—but only after a guitar player leads them in song. "I don't believe in evolution, I know creation's true," they sing, clapping their hands. "Today," the narrator says, "biblical literalism has no more forceful an advocate than Ken Ham." Millions of listeners, we are told, heed "his message that we need to look no further than the Bible to find the truth about who we are." Ham tells his audience: "I believe God created in six literal days, and I believe it's important."

This scene makes an interesting contrast with the scene in Episode One showing Kenneth Miller in a Roman Catholic church. *Evolution* clearly approves of Miller's endorsement of Darwinism, and disapproves of Ham's rejection of it. This also leaves the impression that only fundamentalist Christians reject Darwinism. In fact, some of the strongest critics of Darwin's theory are scientists who happen to be non-fundamentalist Protestants, Catholics, or Jews (as well as agnostics).

We listen to Ham for a few more minutes before the narrator says: "Ken Ham is not the first defender of the faith who is challenging accepted views of science to justify a literal reading of Genesis. Back in 1925, William Jennings Bryan capped his long career as a crusader for Christian values by upholding the State of Tennessee's law banning the teaching of evolution at the famous Scopes monkey trial. Despite a scathing attack on his creationist views, Bryan prevailed."

But this portrayal of William Jennings Bryan is completely false. Bryan did not take biblical chronology literally; instead, he accepted the prevailing scientific view of the age of the Earth. This distortion of history is simply one more attempt to promote the same scientist-vs.-fundamentalist stereotype with which the *Evolution* series began.

The narrator says that anti-evolution efforts following the Scopes trial "had a chilling effect on the teaching of evolution and the publishers of science textbooks. For decades, Darwin seemed to be locked out of America's public schools. But then evolution received an unexpected boost from a very unlikely source—the Soviet Union." When the Soviets launched the first man-made satellite, Sputnik, in 1957, Americans were goaded into action. The narrator continues: "As long-neglected science programs were revived in America's classrooms, evolution was, too. Biblical literalists have been doing their best to discredit Darwin's theory ever since."

This takes the distortion of history one giant step further. It is blatantly false that U.S. science education was "neglected" after the Scopes trial because Darwinism was "locked out of America's public schools." During those supposedly benighted decades, American schools produced more Nobel Prize-winners than the rest of the world put together. And in physiology and medicine—the fields that should have been most stunted by a neglect of Darwinism—the U.S. produced fully twice as many Nobel laureates as all other countries combined.[1]

How about the U.S. space program? Was it harmed by the supposed neglect of Darwinism in public schools? Contrary to what *Evolution* implies, the U.S. space program in 1957 was in good shape. The Soviet Union won the race to launch the first satellite because it had made that one of its highest national priorities. The U.S., on the other hand, had other priorities—such as caring for its citizens and rebuilding a war-torn world. When Sputnik prodded Americans to put more emphasis on space exploration, the U.S. quickly surpassed the Soviet Union and landed men on the Moon. The necessary resources and personnel were already in place; the U.S. didn't have to wait for a new generation of rocket scientists trained in evolution.

The history of 20th-century American science and technology is one of the greatest success stories of all time. *Evolution*'s claim that American science education was "neglected" because of the Scopes trial is completely unjustified. In fact, the claim is so preposterous that it raises serious questions about the integrity of the entire series.

Re-enter Ken Ham, who tells his audience that the biblical flood really happened, and that the fossils we now see were creatures who drowned in the flood and were then buried in rock layers all over the Earth. The scene ends with another rousing song.

B. Controversy at Wheaton College

We drive through a narrow crevice in a mountain as the narrator says: "If you'd been told all your life that the billions of dead things in the Earth got there because of a worldwide flood, the evidence for an ancient Earth comes as a shock."

The driver of a van full of students says: "So we do see evidence of change. But how that change has occurred—whether it has occurred through some sort of a (as Darwin would have said)—some sort of a natural selection, or if it's taken place through some sort of a design—if God has been directly involved in what we see as evolution—that's a bigger question. I think it's a more troubling question for an awful lot of Christians, as well."

The students watch a fossil being excavated as a guide explains that it's about 33 million years old. "At the Wheaton College science station in the Black Hills of South Dakota," the narrator continues, "the shock of the new has started more than one student on his or her way to an understanding of evolutionary history."

Nathan, a geology student, explains how he has struggled to reconcile his belief in the Bible with the scientific evidence: "That's a struggle I've gone through this year. Where is God?" According to the narrator, we are in the eye of a storm: "Wheaton, one of the top fifty schools in America, is committed to exposing its students to the discoveries of science. But as a Christian college, it is also committed to preserving their faith in the God of the Bible."

Nathan describes how as a child he had been indoctrinated in a literal interpretation of Genesis and taught that evolution is evil. We hear from his mother; we attend his local church; and we join his family for a barbecue, where he and his father discuss evolution and the Bible. The son believes Darwinian evolution is true, but his father disagrees.

Back on the Wheaton campus, the narrator continues: "Some of the most troubling questions come, not from science, but from the Bible itself." We meet Emmy, a student of veterinary medicine, who is wrestling with the origin of sin, and the fact that family trees in the Bible all go back to Adam. A group of students sits around a table, trying to reconcile evolution with Christian beliefs about Adam and Eve. The narrator tells us that Wheaton students are free to do this, "but for the professors, open debate on this subject is impossible, thanks to the controversy stirred up by one man's remarks almost forty years ago."

At a Wheaton symposium in 1961, Iowa State University biochemist Walter Hearn said that the same chemical processes that bring each of us into existence today could have produced Adam and Eve. A conservative Christian newspaper spread the word that Wheaton had swallowed evolution wholesale. This was not true, since Hearn had been only one speaker on a diverse panel addressing all aspects of the evolution-creation controversy, but concerned parents and alumni flooded the campus with letters of protest. Wheaton reacted by requiring every faculty member to sign a statement of faith (still in effect today), affirming that all mankind is descended from Adam and Eve, who were created by God.

"Forty years after Walter Hearn shook the campus with his shocking remarks," continues the narrator, "Wheaton is ready to try again." We see Kansas State University geologist Keith B. Miller lecturing to Wheaton students about evolution. Miller explains that he wants to present himself "as a strong advocate for the teaching of evolution and for the centrality of evolution as a unifying scientific theory, and at the same time make very clear my evangelical Christian position."

According to the narrator, "Keith Miller's message to these Christian students is that *all* the evidence, from the ancient fossil record to the latest DNA analysis, compels us to accept the evolutionary theory in full. But for some Wheaton students, the implications of our descent from a common ancestor are still troubling." A student asks Miller how he reconciles evolution with the biblical teaching that we are made in the image of God. He responds: "I personally do not believe that the image of God is connected to our physical appearance, or our origin, as far as how we were brought into being."

Afterwards, Emmy praises Miller for having the courage to discuss his evolutionary beliefs openly. But not everyone on campus is comfortable with Darwin's theory. Peter, an anthropology student, says simply that if he had to choose, he would choose young-earth creationism "just because that's what I grew up with, that's what I'm comfortable with." Beth, a pre-med student, complains of feeling threatened by people who think a "six-day creation is the only way to go." But she still wonders "how God works in us. Where is God's place, if everything does have a natural cause?" Emmy says that she came to Wheaton to "be in a Christian environment where I could *think*."

These are poignant scenes. Children raised in homes where they were taught a literal interpretation of Genesis go off to college, where they are confronted with evidence for an old earth and Darwin's theory of evolution. The ensuing conflicts are very real.

Yet again, however, *Evolution* reinforces the scientist-vs.-fundamentalist stereotype by emphasizing the conflict over biblical literalism, and by leaving us with the impression that once students begin to think they invariably embrace Darwinism. In reality, the conflicts we have witnessed here are only a small part of a much bigger picture. We got a glimpse of the bigger picture from the van driver, who said it has to do with *design*.

From the time of Darwin, the most significant religious objection to his theory focused not on the age of the earth or a literal reading of the Bible, but on his claim that living things are undesigned results of an undirected natural process. It is Darwin's rejection of design and direction—not his challenge to biblical literalism—which has provoked the most controversy among religious believers. By systematically ignoring the bigger picture, *Evolution* distorts the issues and misleads its viewers. We will return to this below.

C. Controversy in a Public High School

As we leave Wheaton, the narrator notes that the faith of some of its students is no longer defined by biblical literalism. "But for Ken Ham," the narrator says, "the frequently repeated fundamentalist expression still holds true: 'God said it; I believe it; that settles it.'" We see Ham conferring in front of a display of toy animals boarding Noah's ark. "Ham and millions of other conservative Christians," the narrator continues, "are convinced that it is the biblical story, not the evolutionary story, that America's children need to hear—not just in Sunday school, but in every school."

According to Ham, "we are concerned about what's happening in high schools. We're concerned about what's happening in the culture. We're concerned that whole generations of children are coming through an educational system basically devoid of the knowledge of God." The scene shifts to a high school corridor crowded with students. Ham continues: "Ultimately, if you're just a mixture of chemicals, what is life all about? Why this sense of hopelessness, this sense of purposelessness? And the reason is because they're given no purpose and meaning in life."

A science teacher gives her students instructions about a computer tutorial. The narrator tells us that this teacher at Jefferson High School in Lafayette, Indiana, is both a scientist and a Christian. She is also "one of thousands of high school science teachers across the country caught in the ongoing struggle between biblical literalism and evolution. The stakes are high—for teachers and students alike." The teacher explains that as a child she accepted the Bible as the word of

God, but as a teenager she found that it conflicted with what she was learning about science. She knew that some of her students were now facing the same conflict, but she was taken aback when over half of the school's students—and 35 members of the faculty—signed a petition demanding the inclusion of "special creation" in the science curriculum.

Her fellow science teacher says he thought the students understood the difference between science and non-science, "and it's fairly obvious to me that if they did at one time, they don't right now." A student then says that her teachers claim not to be accepting or rejecting the existence of God, but when they treat evolution as "the only way" they are indirectly denying God's existence.

A group of students discusses the problem, then the teacher says: "I don't know if this is an isolated incidence of kids just becoming passionate about the situation, or if this is actually the new creationist game-plan: If you can't attack evolution in the Supreme Court, then maybe you can go around and pull one evolution weed at a time to get rid of it. That's what I'm afraid of."

We move to the National Center for Science Education (NCSE) in Oakland, California, an organization that describes itself as "working to defend the teaching of evolution against sectarian attack." According to NCSE's Executive Director Eugenie C. Scott: "People actually don't understand the issues. People are being told, first, you have to choose between faith and science, you have to choose between especially Christianity and evolution. They're being told, Well it's only fair to give both points of view. It's only fair to teach evolution and balance it with creation science or intelligent design theory, or something like that."

Intelligent design theory? Although *Evolution* does its best to portray all critics of Darwin's theory as young-earth biblical literalists—"creation-science" advocates—intelligent design theory is quite different from biblical literalism. Intelligent design theory is based on the hypothesis that some features of living things may be designed. Whether or not a particular feature is designed must be determined on the basis of biological evidence. But the theory says nothing about the Bible. Instead, it includes a critique of the reigning Darwinism—a scientific critique the NCSE does not want students to hear.

Of course, if something is designed it must have a designer. In this sense, intelligent design theory opens the door to the religious realm—a door that Darwinism tries to keep tightly closed. But intelligent design theory by itself makes no claims about the nature of the designer, and scientists currently working within an intelligent design framework include Protestants, Catholics, Jews, agnostics, and others.

Since courts have ruled that creation science cannot be taught in public school science classes, Eugenie Scott and the NCSE lump intelligent design theory with creation science in order to keep it out of science classrooms where it might otherwise be included in discussions of Darwinian evolution. But the differences between intelligent design and creation science are public knowledge; both *The*

Los Angeles Times and *The New York Times* reported on them in 2001. Although *Evolution* claims to be committed to "solid science journalism," it completely ignores these reports. When the *Evolution* series was being made, the producers invited some intelligent design theorists to be interviewed for this last episode. When it became clear that their views would be stereotyped as a form of religious fundamentalism, however, the intelligent design theorists refused to take part.[2]

Scott continues: "Evolution—or science in general—can't say anything about whether God did or did not have anything to do with it. All evolution as a science can tell us is what happens. Can't tell us whodunit. And as [for] what happened, the evidence is extremely strong that the galaxies evolved, the planets evolved, the sun evolved, and living things on Earth shared common ancestors."

But this last statement mixes apples and oranges. To say that galaxies, planets and the sun evolved is merely to say that they changed over time. To say that all living things evolved from common ancestors makes a much more specific claim. The evidence for the former may be "extremely strong," but where is the evidence for the latter? Despite *Evolution*'s promise to show us the "underlying evidence" for evolutionary theory, it has presented almost no evidence for the common ancestry claim. One key piece of evidence—the supposed universality of the genetic code—even turned out to be false.

Furthermore, if this series is any indication, evolution has a ***lot*** to say about "whether God did or did not have anything to do with it." In Episode One, Stephen Jay Gould pooh-poohed the idea that "God had several independent lineages and they were all moving in certain pre-ordained directions which pleased His sense of how a uniform and harmonious world ought to be put together." In the same episode, Kenneth Miller argued that the vertebrate eye was not designed by God, but produced by evolution. And in Episode Five, Geoffrey Miller assured us that "it wasn't God, it was our ancestors" that produced the modern human brain by "choosing their sexual partners."

The camera focuses on colored pins stuck into a large wall map of the United States. "Calls come in from across America," says the narrator, "from teachers who continue to be accused of locking God out of their classrooms." Among the teachers who contact the NCSE are the ones in Lafayette, Indiana.

D. The Lafayette School Board

Jefferson High School students carry their petition to the Lafayette School Board, which listens politely to their statements. One student emphasizes that "those of us supporting this petition do not advocate the banning of teaching of the theory of evolution; however, we believe that the theory of evolution should be taught alongside the alternative theory of special creation. Let us be taught the facts, so that we can decide on our own."

According to the narrator: "For these students, the argument isn't about science versus the Bible; it's about which views of science will be taught. It is a tactic pioneered in 1961, when a revolutionary book by Henry Morris and John Whitcomb used carefully selected scientific evidence to support the creationist cause." Scott adds: "*The Genesis Flood* is the foundational document for creation science. Everything else has been built upon this book."

The narrator describes a 1981 Louisiana law requiring that creation science be taught alongside evolution science, and how the U.S. Supreme Court overturned the law in 1987 on the grounds that it "violated the First Amendment separation of church and state"—though the Supreme Court also ruled that "alternatives to evolutionary theory can be taught if they have a scientific basis."

"Of course teachers have a right to teach any and all scientific views about the origin of humans or any other scientific theory," Scott says, emphasizing the word "scientific." But "one reason why the creationists have worked so hard to try to present their ideas as being scientific is so they can duck under the First Amendment."

So Scott is opposed to presenting views in science classrooms that are not scientific. As we saw in Episode Five, however, even many evolutionary biologists consider evolutionary psychology to be unscientific. And as we saw in Episode Six, even many evolutionary biologists consider memetic evolution to be unscientific. Why doesn't Scott oppose the teaching of *these* views? Why does she support using this series, for example, as a teaching instrument in public schools?

We return to Lafayette, Indiana. The students want to learn the facts so they can decide for themselves. One of the science teachers feels the students don't understand the nature of science, because "creation and any Supreme Being can't be addressed in a science classroom." Another science teacher says: "In science, ideas are supported by evidence, and that evidence has to be peer-reviewed, and it has to be repeatable, and it has to be testable. And creationism is *not* that." The first teacher lays the blame partly on the students' parents, who (she says) don't want them even to hear about evolution.

The Lafayette School Board hears the students out, but decides to deny their petition on the grounds that biological science is clearly defined, and special creation does not fall within that definition. "The decision preserved the integrity of Jefferson High's science curriculum," the narrator says, "but the teachers know this is not the end of the debate."

The teacher we first met at Jefferson High remarks: "I have yet to hear of a case where they've given equal time in a science classroom; however, I have heard of cases where they've removed evolution from the curriculum. And I don't think the three of us would have continued teaching here had that been the case. I can't speak for them, but I really don't think as an educator I could teach biology and do it well, if I couldn't talk about the natural processes that make it work. To take that element out would be removing one of the—well *the* major pillar that supports that whole field of science."

But the students petitioned their school board to include "the facts, so that we can decide on our own." They specifically said they did *not* want evolution taken out. Why, then, does this scene conclude with a teacher expressing concern over the danger of removing evolution from the curriculum? That happened in Tennessee in 1925. The Scopes trial and Walter Hearn's experience show us that Christians have sometimes censored evolution. What just happened in Lafayette, however, was the exact opposite: Darwinian evolution was granted exclusive dominion over the science classroom, and all discussions of special creation— including any facts that might support it—were banned.

Whatever one may think of special creation, there is no doubt that *Evolution* is spinning this story to make the victim look like the bad guy. In Lafayette, special creation was the censored, not the censor. And the censorship continues: On August 14, 2001, the Lafayette *Journal and Courier* reported that a Jefferson High School science teacher had been officially reprimanded by the district superintendent just for mentioning creation in his classroom.

Darwinian censorship is frequently used not only to ban discussions of creation, but also to block all criticism of Darwin's theory. In 2000 and 2001 Roger DeHart, a high school biology teacher in Burlington, Washington, was prohibited by his superintendent from giving students an article by evolutionist Stephen Jay Gould, which pointed out that some of the evidence for evolution in their textbook had been faked! And when William Dembski, director of a research institute at Baylor University in Texas, organized an international conference in 2000 that brought together critics as well as defenders of Darwin's theory, he encountered a storm of opposition and was eventually removed from his position.

By including the Lafayette School Board story with its misleading spin, *Evolution* may be trying to influence the political decisions of local school boards. In an internal memo dated June 15, 2001, *Evolution*'s producers announced their plan to "co-opt existing local dialogue about teaching evolution in schools." The "goal of *Evolution*," they wrote, is to "promote participation," and one way to do that is "getting involved in local school boards." It seems that this story about the Lafayette School Board is part of a strategy to use public television to influence elected officials.[3]

E. This View of Life

We return to Wheaton. A college spokesman says: "Are we placing students' faith at risk by examining these hard questions? Absolutely. But I would add, additionally, that there is no such thing as a safe place from which to hide from these issues. If we engage in the most rigid biblical literalism, the fact that our students live in a real world indicates that their faith is always at risk. Christians believe that our faith is rooted in real happenings in a real world, and so to try and structure a place or a way of conceptualizing our faith that insulates us and iso-lates us from risk is to rob Christianity of its very essence."

Emmy, the veterinary medicine student, says she doesn't want to come across as a religious fanatic. "I want to be educated, I want to be intelligent, I want to have answers." Beth, the pre-med student, says: "Because we look for natural causes in things doesn't mean we think that that's all there is. It doesn't mean that we're throwing out the meaning of life. We're just studying what God has made, however He made it." And now that Nathan has accepted evolutionary theory, he finds that he has the "freedom to say, 'Wow, God is bigger than the box that I may have put Him in.'"

Except for Peter, the anthropology student who remains a biblical literalist mainly because he grew up with it, all the Wheaton students we have met think that biblical literalism is for the ignorant and narrow-minded, while evolution is for the educated and broad-minded. Presumably, we are expected to conclude that skepticism about evolution naturally disappears as people grow up and get edu-cated. One would never guess that a growing number of highly educated scien-tists—as we saw above—are becoming increasingly skeptical of evolutionary theory. "Keith Miller's message to these Christian students," we are told, "is that *all* the evidence, from the ancient fossil record to the latest DNA analysis, com-pels us to accept the evolutionary theory in full." That's a very strong claim—a claim that many scientists would question. Are we supposed to believe that the only people at Wheaton who had a problem with it were the biblical literalists?

Actually, biblical literalists are *not* the only people who disbelieve in Darwin-ian evolution. Over the past decades, Gallup polls have consistently shown that roughly 45% of the American people believe that God created the world in its present form only a few thousand years ago. These are the biblical literalists that this series portrays as the only critics of Darwinian evolution. Another 45% or so believe that things have changed over a long period of time, but that God guided the process. This might be called "evolution" in the broad sense of "change over time," but it is certainly not Darwinian evolution. Only about 10% of Americans subscribe to Darwin's theory that all living things—including us—are undesigned results of undirected natural processes. So Darwinian evolution is actually embraced by only a small minority of the American people.

Why didn't *Evolution* interview Huston Smith, who is probably the most highly regarded living authority on the world's religions? According to Smith, Darwinism has been a major factor in "the modern loss of faith in transcendence, basic to the traditional/religious worldview." Nothing here about biblical literal-ism—or even Christianity. Smith is talking about *all* of the world's major reli-gions. Like Daniel Dennett, Huston Smith sees Darwinism as corrosive to the faith in transcendence that lies at the root of all religion. But while Dennett considers Darwinism to be true, Smith is a vocal critic of it. Among other things, Smith maintains that Darwinism is "supported more by atheistic philosophical assumptions than by scientific evidence."[4]

So out of the vast spectrum of the world's religious beliefs, *Evolution* gives voice only to biblical literalists—whom it dismisses as uneducated and doctrinaire—and to the small minority of Christians who subscribe to Darwin's theory. The series completely ignores the hundreds of millions of other Christians—not to mention Muslims, Hindus, and orthodox Jews—who reject the Darwinian doctrine that all living things—including us—are undesigned results of undirected natural processes. We have seen how shallow and lopsided *Evolution* can be in its presentation of controversies among evolutionary biologists. But its presentation of the evolution-creation controversy is even worse.

As the sun sets over the Pacific, the narrator brings the eight-hour series to a fitting close, quoting from the conclusion of Darwin's *The Origin of Species:* "There is grandeur in this view of life, with its several powers, having been originally breathed by the Creator into a few forms or into one; . . . from so simple a beginning endless forms most beautiful . . . have been, and are being, evolved."

Evolution began with the Bible, and now it ends with the Creator. Despite the producers' assurance that they would avoid "the religious realm," *Evolution* has had a great deal to say about it. The first episode dealt with religion extensively, Episodes Two and Six touched on it briefly, Episode Five mentioned it repeatedly, and this final episode was devoted to it entirely. Far from avoiding it, *Evolution* has spoken to the religious realm from start to finish.

And what did it say about religion? The message is unmistakable. As far as *Evolution* is concerned, it's OK for people to believe in God, as long as their beliefs don't conflict with Darwinian evolution. A religion that fully accepts Darwin's theory is good. All others are bad.

Notes

1. For more information about Ken Ham's views, go to:

    http://www.answersingenesis.org/home.asp

The usual stereotype of the Scopes trial comes, not from the 1925 trial itself, but from the 1960 motion picture, "Inherit the Wind." The differences between the two are described in Edward J. Larson's Pulitzer Prize-winning book, *Summer for the Gods: The Scopes Trial and America's Continuing Debate over Science and Religion* (New York: Basic Books, 1997).

The years during which *Evolution* claims U.S. public science education was "neglected" due to censorship of Darwinian evolution extended from 1925 (the year of the Scopes trial) to 1957 (the year of Sputnik). There would have been a slight delay in the effect of the Scopes trial on high school students—the first graduating class after the trial was Spring 1926, and the claimed effect would presumably have increased thereafter; so the thirty years from 1927 to 1957 are the crucial ones. A sampling of twentieth-century U.S. Nobel science laureates shows ages ranging from 30s to 70s, with an average age in the mid-50s. A 55-year-old would have gone through high school about four decades years earlier; so high school students from the period 1927-1957 would, on average, have won Nobel Prizes from 1967-1997.

Between 1967 and 1997, prizes were awarded as follows:

Category	United States	All other countries
Physics	40	28
Chemistry	30	27
Physiology/ Medicine	48	24
Total	**118**	**79**

Note that in physiology and medicine, the fields (according to *Evolution*) most likely to be adversely affected by neglecting Darwin's theory, U.S. scientists won twice as many Nobel Prizes during this period as all other countries put together.

2. For more information about the NCSE, go to:

http://www.natcenscied.org/

For stories about the controversy in Lafayette, Indiana, consult the local newspaper, the *Journal-Courier* (stories are archived for 14 days) at:

http://www.jconline.com/.

Some prominent intelligent design theorists are Baylor University mathematician William A. Dembski, author of *The Design Inference* (Cambridge: Cambridge University Press, 1998) and *Intelligent Design* (Downers Grove, IL: InterVarsity Press, 1999); Lehigh University biochemist Michael J. Behe, author of *Darwin's Black Box: The Biochemical Challenge to Evolution* (New York: The Free Press, 1996); and University of Otago molecular biologist Michael Denton, author of *Evolution: A Theory in Crisis* (Bethesda, MD: Adler & Adler, 1986) and *Nature's Destiny: How the Laws of Biology Reveal Purpose in the Universe* (New York: The Free Press, 1998). A good recent anthology of writings on intelligent design for lay people is William A. Dembski and James M. Kushiner (editors), *Signs of Intelligence: Understanding Intelligent Design* (Grand Rapids, MI: Brazos Press, 2001). For more information on intelligent design theory, go to:

http://www.discovery.org/crsc/

Also see the following: Access Research Network, Frequently Asked Questions about Intelligent Design:

http://www.arn.org/id_faq.htm

Mark Hartwig, "The World of Design," *Teachers in Focus* (September, 2000):

http://www.arn.org/docs/hartwig/mh_worldofdesign.htm.

For a recent journalistic report on intelligent design theory, see Teresa Watanabe,

"Enlisting Science to Find the Fingerprints of a Creator," *The Los Angeles Times* (March 25, 2001), 1. Watanabe wrote: "Unlike biblical literalists who believe God created the world in six days, most theorists of intelligent design are reputable university scholars who accept evolution to a point. But they question whether Darwinist mechanisms of random mutation and natural selection can fully account for life's astonishing complexity. Instead, using arguments ranging from biochemistry to probability theory, they posit that some sort of intelligence prompted the unfolding of life—say, by producing the information code in the DNA."

See also James Glanz, "Darwin vs. Design: Evolutionists' New Battle," *The New York Times* (April 8, 2001), 1. Glanz wrote: "Evolutionists find themselves arrayed not against traditional creationism, with its roots in biblical literalism, but against a more sophisticated idea: the intelligent design theory. Proponents of this theory, led by a group of academics and intellectuals and including some biblical creationists, accept that the earth is billions of years old, not the thousands of years suggested by a literal reading of the Bible. But they dispute the idea that natural selection, the force Darwin suggested drove evolution, is enough to explain the complexity of the earth's plants and animals. That complexity, they say, must be the work of an intelligent designer."

3. For more information on the censorship of Roger DeHart, go to:

 http://www.arn.org/docs/pearcey/np_world-creationmythology62400.htm

 http://www.discovery.org/news/someTeachersTalkAlternativ.html

For more information on the Baylor controversy surrounding William Dembski, go to:

 http://www.arn.org/docs/dembski/wd_dallasobserver0101.htm

 http://www.touchstonemag.com/docs/issues/14.4docs/14-4pg54.html

 http://www.fsf.vu.lt/filk/mps/Being%20Methodologically%20Correct.htm

On June 15, 2001, the producers of *Evolution* distributed an internal memo to PBS affiliates entitled "The Evolution Controversy: Use It Or Lose It." Among other things, the memo listed under "Key Evolution Marketing" several Project Outreach goals, one of which was to "co-opt existing local dialogue about teaching evolution in schools." Under "Project Messaging," the memo listed "the six most important messages we can convey." One of these was: "The goal of *Evolution* is to create a dialogue and promote participation. . . . Participation can occur in many ways: watching the TV series, logging on the Web site, helping with kids' science homework, getting involved in school board meetings, cleaning up your local environment, and countless other activities that further science literacy and our understanding of the natural world."

4. The quotations from Huston Smith are from "Huston Smith Replies to Barbour, Goodenough, and Peterson," *Zygon* 36, No. 2 (June, 2001), 223-231. See also Huston Smith, *Why Religion Matters: The Fate of the Human Spirit in an Age of Disbelief* (New York: Harper Collins, 2001). Huston Smith is the author of *The World's Religions* (New York: Harper, 1992).

Conclusion

The show's over. What did we learn from it?

A. Underlying Evidence for Evolution?

The producers promised to provide us with "underlying evidence behind claims of fact and proposed theories." We saw lots of data, but how much of it was evidence for Darwin's theory?

We saw lots of data from the fossil record. We saw fossils of some of the first animals (from the Cambrian explosion). We saw fossils of early land animals, dinosaurs, early mammals, whales, ape-like creatures, and humans. Clearly, the composition of the Earth's biosphere has changed over time. Some things that used to inhabit the Earth are no longer with us, and some things we see around us were not always here.

But the fossil record does not—and cannot—show us ancestry and descent. Maybe some of the fossils we saw were ancestral to others, and maybe they weren't. As Henry Gee, chief science writer for *Nature,* wrote in 1999, "the intervals of time that separate fossils are so huge that we cannot say anything definite about their possible connection through ancestry and descent." According to Gee: "To take a line of fossils and claim that they represent a lineage is not a scientific hypothesis that can be tested, but an assertion that carries the same validity as a bedtime story—amusing, perhaps even instructive, but not scientific."[1]

Every time someone refers to fossils as ancestors, that person is *assuming* that Darwin's theory of common ancestry is true, and then stringing fossils together in chains of ancestry and descent. But how do we know that Darwin's theory is true?

We saw similarities and differences among fossils, and between fossils and living species, and among living species. But many of these were known to Darwin's predecessors, who attributed them to designed construction rather than to unguided evolution. In particular, we saw lots of similarities and differences between humans and chimpanzees; by themselves, however, such features tell us nothing about our ancestry.

How can we know that similarities are due to common ancestry rather than common design? Only by showing that natural processes can produce them. Otherwise, the possibility remains that—like automobiles—living things were constructed by design.

If we could observe the process of descent with modification, that would settle the matter; but we can't. So an evolutionary biologist begins by *assuming* that Darwin was right, and interprets similarities and differences from that perspective. But how do we know that Darwin was right?

We saw data showing that some organisms—mainly viruses and bacteria—change over time. But the changes we saw were all *within* species. If we started with HIV, we finished with HIV. The TB in an Egyptian mummy was the same species found in New York City. Changes within species were well known to Darwin's predecessors, but they do not provide evidence for his theory about the origin of *new* species.

Once more, if we simply *assume* the truth of Darwin's theory, then it's easy to imagine one species changing into another. But why should we assume—in the absence of evidence—that Darwin's theory is true?

It may seem uncharitable to persist in questioning the truth of Darwin's theory, and to keep insisting that *Evolution* show us evidence that actually supports it. But that's what we were promised. And that's what science is all about.

So the data we have reviewed so far are not the kind that supports Darwin's theory. We were also shown another kind of data in *Evolution*—the kind that appears to support the theory, but isn't true.

B. Distortions of Scientific and Historical Facts

We were told that the genetic code—the "language" by which DNA specifies protein sequences—is the same in all living things, and that this "is powerful evidence that they all evolved on a single tree of life." But molecular biologists have known for years that the genetic code is *not* the same in all living things. What we were told is false.

Then we were told that HIV takes only "minutes to hours to move from one species to another." This could provide some of the evidence that Darwin's theory needs. But no new species formed. The claim is false.

Then we were told that a tiny handful of powerful genes such as *Antennapedia*—which when mutated causes flies to sprout legs from their heads—are the "architects of the body" and the "genetic engine of evolution." But these genes don't exert their effects until *after* an animal's body is formed, so something else must be the body's architect. And the fact that they are similar in radically different animals, and that mutations in them never make animals more fit, shows that such genes cannot explain the evolution of one kind of animal from another.

Finally, we were told about "people" who lived millions of years ago—though we were also told that people (as we normally use the word) first appeared about fifty thousand years ago. *Evolution* called the genetic code universal when it isn't, said HIV moved to a new species when it didn't, claimed genes like *Antennapedia* make animal bodies when they can't, and called ape-like creatures people when they weren't. In other words, *Evolution* systematically misrepresented the evidence to make it look as though it supports Darwin's theory—when it doesn't.

Evolution also distorted historical facts as well as scientific ones. It mischaracterized the details of Darwin's life to promote the scientist-vs.-fundamentalist stereotype that runs throughout the series. In fact, much of the opposition to Darwin's theory during his lifetime came from scientists, not theologians. And among the theologians, criticism was aimed primarily at Darwin's rejection of design, not his challenge to biblical literalism.

The situation is similar today. Although *Evolution* would have us believe that all of the opposition to Darwin's theory comes from biblical literalists like Ken Ham, only about 10% of Americans accept Darwinian evolution in full. The vast majority of Americans—not just biblical literalists—have a problem with Darwin's claim that living things are undesigned products of an unguided process.

Even worse, *Evolution* completely ignored or misrepresented the growing number of highly qualified scientists who criticize or reject Darwin's theory. For example, the series dismissed intelligent design theorists as biblical literalists, even though articles by mainstream journalists in *The Los Angeles Times* and *The New York Times* have pointed out that this is false.

Evolution's most egregious distortions of history, however, involve the 1925 Scopes trial and its aftermath. First, it misrepresented William Jennings Bryan as a biblical literalist in order to promote the scientist-vs.-fundamentalist stereotype. Second, it made the preposterous claim that U.S. science education was "neglected" in the decades following the trial—the same decades when America produced more Nobel laureates than the rest of the world combined. And finally, it portrayed modern evolutionists as the victims of Scopes-style censorship, when in fact the situation is now exactly the reverse, with Darwinists censoring their critics.

So *Evolution* presented us with some data that didn't really support Darwin's theory, and some data that appeared to support the theory but turned out to be false. The series also distorted history to discredit Darwin's critics. Perhaps even more surprising, however, was the way *Evolution* ignored scientists who accept Darwin's theory but who disagree with many of the things this series said about it.

C. Shallow and Lopsided Coverage of Scientific Controversies

According to its producers, one of *Evolution*'s goals was to report on "areas where the science is sound." Yet many of the areas covered by the series are far from being sound—in fact, they are highly controversial even among evolutionary biologists.

We were told that sexual reproduction exists because it generates genetic variability. This supposedly enables members of sexually reproducing species to resist parasites and adapt to changing environments. Although many biologists believe this, the evidence is inconclusive, and the issue remains highly controversial. As *Science* reported in 1998, biologists "haven't solved the mystery of sex yet," partly because of "extremely lousy experimental data." So "how sex began and why it thrived remain a mystery."

We watched a long interview with evolutionary psychologist Geoffrey Miller, who thinks that the human brain is like a peacock's tail. Both, he thinks, are products of sexual selection. Miller also regards all of human culture as a by-product of sexual choices. But many biologists regard evolutionary psychology as a non-science. According to University of Chicago evolutionary biologist Jerry Coyne, evolutionary psychologists "deal in their own dogmas, and not in propositions of science." And American Museum of Natural History paleoanthropologist Ian Tattersall wrote of Miller's work: "In the end we are looking here at a product of the storyteller's art, not of science."

Finally, we heard a lot about memes from Richard Dawkins and Susan Blackmore—but not a word from the many biologists who consider the concept hopelessly vague and unscientific. Stephen Jay Gould, whom we met in Episode One, considers it a "meaningless metaphor." And Jerry Coyne considers memes "but a flashy new wrapping around a parcel of old and conventional ideas."[2]

Evolution's coverage of these highly controversial areas was completely one-sided. No critical voices were raised, leaving viewers with the misleading impression that these were "areas where the science is sound." Instead of educating viewers about how science really works, *Evolution* chose to indoctrinate them in the latest fads.

And two of these—evolutionary psychology and memetic evolution—also served the producers' goal of convincing us that almost everything revolves around Darwinian evolution.

D. Is Evolution Indispensable to Medicine, Agriculture, and Choice of Mate?

One goal of the series, the producers said, was to show that evolution is important to "almost every aspect of human life," from "medicine to agriculture to a person's choice of mate." Let's see how well it did.

Episodes One and Four suggested that evolution is important to medicine because it has been instrumental in treating HIV patients. In the first case, patients were taken off of drugs so the HIV in their bodies could lose its drug-resistance. This strategy of interrupting treatment seemed to work, but how much did it owe to evolution? Contrary to what we were told, no new species of HIV emerged. The change that was observed could have been predicted from principles of artificial selection that were known for centuries before Darwin. In Episode Four, the connection between HIV and evolution was even more tenuous, and the discovery of a protective mutation did nothing to help patients.

The story of multi-drug resistance in tuberculosis was interesting and important, but like the HIV story in Episode One it owed nothing to Darwin and did not provide evidence for his theory. And the main lesson from the cholera story was that we should drink clean water—but we already knew that.

The story of insulin in Episode Six was the most far-fetched of all. Effective treatment of diabetes is a triumph of modern medicine, but calling it a "meme" and putting a Darwinian "spin" on it is sheer nonsense. This—like Blackmore's assertion that memes invented the Internet—sounds suspiciously like the Soviets' old insistence that they invented the telephone.

The leafy spurge story in Episode Three was *Evolution*'s feeble bid to lay claim to agriculture. But using an insect to control an agricultural pest is nothing new. The ancient Chinese and Yemenis did it, with no help from Darwin. The leafy spurge story was interesting, but it neither depended on evolutionary theory nor provided any support for it.

Finally, the advice that *Evolution* gave us in choosing a mate was perhaps the least useful of all. Choose your mate by smelling his T-shirt, or by selecting his face from a computer line-up, or because he has a bigger brain. Or choose memetic evolution and do whatever you want, without regard for biological consequences.

It seems, then, that Darwinian evolution isn't really important to medicine, agriculture, or match-making after all. Nevertheless, the producers of the series make it clear that there is one realm where it is absolutely essential: religion.

E. The Religious Realm

In keeping with "solid science journalism," the producers of *Evolution* set out to examine "empirically-testable explanations" while avoiding "the religious realm."

Yet *Evolution* dealt with the religious realm from start to finish. It twisted historical facts to make critics of evolution look like biblical literalists. It employed religious symbolism such as Michelangelo's painting of God touching Adam to convey its message that humans are not special. And it quoted—repeatedly and approvingly—anti-religious statements by a whole parade of Darwinists.

In Episode One, Daniel Dennett told us that after Darwin we no longer have "meaning coming from on high and being ordained from the top down." Stephen Jay Gould pooh-poohed the idea that "God had several independent lineages and they were all moving in certain pre-ordained directions which pleased His sense of how a uniform and harmonious world ought to be put together." And James Moore stated the problem Emma Darwin had with her husband's theory: "Now if nature, by itself, unaided by God, could make an eye, then what else *couldn't* nature do? Nature could do anything!"

Kenneth Miller argued that imperfections in the vertebrate eye were "proof" that it was due to evolution rather than God's design. We visited him in church, and he told us: "I'm an orthodox Catholic and I'm an orthodox Darwinist." Then he said that "if God is working today in concert with the laws of nature, with physical laws and so forth, He probably worked in concert with them in the past. In a sense, in a sense, He's the guy who made up the rules of the game, and He manages to act within those rules." Finally, James Moore concluded Episode One by assuring us that "Darwin's vision of nature was, I believe, fundamentally a religious vision."

Episode Five taught us that Darwin replaced the idea of God creating ornate feathers with his theory of sexual selection. Geoffrey Miller said "it wasn't God, it was our ancestors . . . choosing their sexual partners" that accounted for the origin of the human brain. Then Handel's *Messiah* was used to illustrate Miller's claim that all of human culture is a result of our sexual instincts.

Finally, Episode Seven was devoted *entirely* to religion. We watched biblical literalist Ken Ham lecturing about creation in a church; we witnessed students at Wheaton College struggling with their Christian upbringing; and we saw a local school board deny a student petition to teach creation alongside evolution. Yet we saw and heard nothing from critics of Darwinian evolution—either scientific or religious—who are *not* biblical literalists. The message was clear: religion is OK in its place, as long as it doesn't challenge Darwinism.

So *Evolution* had quite a lot to say about the religious realm. And far from reporting objectively on the wide range of religious viewpoints in America, it singled out only two—one of which it obviously preferred over the other. Now, the producers of *Evolution* are entitled to their opinion. In America, everyone is. But the government, and other public resources such as PBS, are not supposed to favor one religion over another. What is the justification for broadcasting this series on public television, and distributing it to public schools, when it is so clearly biased, both scientifically and religiously?

F. Evolution and Public Policy

PBS is funded in part by American taxpayers. It is thus supposed to remain neutral in religious matters. It is absolutely inappropriate for PBS to engage in activities that promote one religious view over another.

It is also inappropriate for PBS to attempt to influence the political process. Yet the producers of the *Evolution* series are trying to do just that. According to the June 15, 2001, document cited in the introduction, one goal of the project is to "co-opt existing local dialogue about teaching evolution in schools." Another goal is to "promote participation," including "getting involved with local school boards." Moreover, "government officials" are identified as one of the target audiences for the series, and the publicity campaign accompanying the series will include the writing of op-eds. Clearly, one purpose of *Evolution* is to influence school boards and to promote political action regarding how evolution is taught in public schools.

The political agenda behind *Evolution* is made even more explicit by its enlistment of Eugenie Scott as an official spokesperson for the project. As we have seen, Scott is the Executive Director the National Center for Science Education (NCSE), an advocacy group that by its own description is dedicated to "defending the teaching of evolution in the public schools." As we have also seen, the NCSE routinely lumps together all critics of Darwinism as "creationists." According to the group's web site, the NCSE provides "expert testimony for school board hearings," supplies citizens with "advice on how to organize" when "faced with local creationist challenges," and assists legal organizations that litigate "evolution/creation cases." It is a single-issue group that promotes one side in the political debate over evolution in public education. It is therefore completely inappropriate for PBS to enlist NCSE's executive director as an official spokesperson on this project—while excluding other views.[3]

The American people—and especially America's students—deserve to be informed about the controversy over Darwin's theory of evolution. But the PBS *Evolution* series is not a sincere effort to inform. Instead, it is an effort to make Darwinian evolution seem more scientific than it really is, to promote one religious viewpoint over others, and to influence local school boards to grant exclusive control to a controversial theory. This is not education. This is not good science journalism. This is propaganda.

Notes

1. Henry Gee, *In Search of Deep Time* (New York: The Free Press, 1999), 23, 32, 113-117, 202.

2. The quotations from the 1998 special issue of *Science* are from Bernice Wuethrich, "Why Sex? Putting Theory to the Test," *Science* 281 (1998), 1980-1982. The same issue included the following articles of interest: Pamela Hines & Elizabeth Culotta, "The Evolution of Sex," *Science* 281 (1998), 1979; N. H. Barton & B. Charlesworth, "Why Sex and Recombination?" *Science* 281 (1998), 1986-1990.

   The Coyne quotation is from Jerry A. Coyne, "Of Vice and Men: The fairy tales of evolutionary psychology," a review of Randy Thornhill and Craig Palmer's *A Natural History of Rape,* in *The New Republic* (April 3, 2000), last page. The entire review is available at:

   http://www.thenewrepublic.com/040300/coyne040300.html

   The Tattersall quotation is from Ian Tattersall, "Whatever turns you on," a review of Geoffrey Miller's book, *The Mating Mind,* in the *New York Times Book Review* (June 11, 2000).

Stephen Jay Gould called memes a "meaningless metaphor" on a radio show November 11, 1996. See Susan Blackmore, "Memes, Minds and Selves," at:

http://www.tribunes.com/tribune/art98/blac.htm#b.

The Coyne quotation is from Jerry A. Coyne, "The self-centred meme," a review of Susan Blackmore's *The Meme Machine,* in *Nature* (April 29, 1999), 767-768.

3. Quotations from the producers about their goals are taken from "The Evolution Controversy: Use It Or Lose It."—a document prepared by Evolution Project/WGBH Boston and distributed to PBS affiliates on June 15, 2001. The document concludes by suggesting that "any further questions" should be directed to WGBH, giving the following information:

The web site for WGBH is http://www.wgbh.org/

Related web sites include: http://www.pbs.org/wgbh/nova/

http://www.pbs.org/

Other contact information:

WGBH

125 Western Avenue

Boston, MA 02134

(617) 300-2000

(617) 300-5400

For more information about the National Center for Science Education (NCSE), go to http://www.natcenscied.org/

ACTIVITY

1

Who Were Darwin's Critics?

Accompanies the *Viewer's Guide*, Chapter 1-A, "The Voyage of the Beagle"

Overview

Virtually every scientific revolution has had its critics. The Darwinian revolution was no exception. Indeed, Darwin's theory of evolution—in its contemporary form—still has many vocal critics. For a variety of historical and cultural reasons, the controversy has often been cast as a battle between science and biblical literalism. But Darwin's theory was opposed not only by biblical literalists, but also by a wide variety of religious believers and scientists. In this research project, students will learn about some of Darwin's religious and scientific critics, and why they objected to Darwin's theory.

Learning Objectives

- Students will be aware that Darwin's theory drew criticism from diverse segments of his society.

- Students will be aware of some of the historical works about Darwin's life and theory.

- Students will be able to identify some of Darwin's scientific and religious critics, as well as some of the critics' objections to Darwin's theory.

Directions

Have students discuss the scenes about Darwin and Fitzroy. Ask the following questions:

1. What did you think of the scenes that showed Darwin and Fitzroy?

2. Was Fitzroy typical of most people who criticized Darwin?

3. Can you think of any other people during Darwin's time who criticized his views?

Hand out the worksheet on the following page and go over its directions with students.

Who Were Darwin's Critics?

Directions

The following people were critical of Darwin's theory of evolution:

Louis Agassiz	Karl Ernst von Baer	John Herschel
Charles Hodge	Fleeming Jenkins	Charles Lyell
John Stuart Mill	St. George Jackson Mivart	Richard Owen
Adam Sedgwick (geologist)	William Thomson (Lord Kelvin)	William Whewell
Samuel Wilberforce		

- Choose five people from the list above and answer the following questions about each:
  1. Who is he?
  2. What was his occupation?
  3. What is he best known for?
  4. What were his objections to Darwin's theory?

- Keep a log of how opponents of evolution are portrayed in Episodes Two through Seven. It doesn't have to be extensive. How does this compare to the diversity of people opposing Darwin's theory during his own time?

Sources

You can find information about these people on the Web and in a number of printed books and articles. Here are a few places to look.

On the Web

Charles Darwin. *The Autobiography of Charles Darwin* (as published in *The Life and Letters of Charles Darwin*), edited by Francis Darwin.

  http://digital.library.upenn.edu/webbin/gutbook/lookup?num=2010

Charles Darwin. *The Life and Letters of Charles Darwin,* edited by Francis Darwin.

  Volume I: http://digital.library.upenn.edu/webbin/gutbook/lookup?num=2087

  Volume II: http://digital.library.upenn.edu/webbin/gutbook/lookup?num=2088

Charles Darwin. *More Letters of Charles Darwin,* edited by Francis Darwin and A. C. Seward:

  Volume I: http://digital.library.upenn.edu/webbin/gutbook/lookup?num=2739

Volume II: http://digital.library.upenn.edu/webbin/gutbook/lookup?num=2740

Charles Darwin. *On the Origin of Species,* 6th ed.

http://digital.library.upenn.edu/webbin/gutbook/lookup?num=2009

Other Web sources can be found with the Google search engine (www.google.com). To find Web sites that discuss these people's views of Darwin and his theory, use their names and such keywords as "Darwin" or "evolution" as your search terms. Remember, though, Web pages can vary in their reliability and accuracy. Which pages do you think are most reliable and why?

Printed Sources

Peter J. Bowler. *Evolution: The History of an Idea.* 2d ed. Berkeley: University of California Press, 1989.

E. Janet Browne. *Charles Darwin: A Biography.* New York: Knopf, 1995.

Adrian Desmond and James Moore. *Darwin: The Life of a Tormented Evolutionist.* New York: W.W. Norton & Company, 1991.

Loren Eisley. *Darwin's Century: Evolution and the Men Who Discovered It.* New York: Doubleday, 1958.

Gertrude Himmelfarb. *Darwin and the Darwinian Revolution.* Garden City, NY: Anchor Books, 1962.

David Hull. *Darwin and His Critics: The Reception of Darwin's Theory of Evolution by the Scientific Community.* Chicago: University of Chicago Press, 1983.

James R. Moore. *The Post-Darwinian Controversies.* Cambridge: Cambridge University Press, 1979.

2

Darwin and His Finches

Accompanies the *Viewer's Guide,* Chapter 1-C, "The Legend of 'Darwin's Finches'"

Overview

The process of discovery in science is seldom ever neat. Sometimes an idea comes in a sudden flash of insight. Other times it may develop over a long time. Still other times, a scientist may miss an insight because he was looking in the wrong places or because there was a flaw in the way he gathered his data. Unfortunately, popular accounts of science tend to overlook the very human way that even great scientists went about their work. In the pursuit of a compelling story, the storytellers often sweep away all the messiness that can go with real science. Instead of scientists, we are shown heroes and legends.

This is no less true of Darwin than anyone else. It is commonly believed that Darwin's insight about evolution was sparked by the finches of the Galápagos Islands. But is this how it happened? In this activity, students will read through relevant chapters of *The Voyage of the Beagle* and *The Origin of Species* by Charles Darwin, and *Darwin: The Life of a Tormented Evolutionist,* by science historians Adrian Desmond and James Moore, to determine what helped give Darwin his idea.

Note: This activity has been structured as a research project. If you wish to adapt it to a single class session, you can provide excerpts from Darwin's work and brief quotes from Desmond's and Moore's book.

Learning Objectives

- Students will understand that scientific discovery is sometimes a messy process.

- Students will understand that popular accounts of history and science may conflict with scholarly accounts and with the evidence.

- Students will synthesize material from various sources to determine what factors may have inspired Darwin's theory of evolution.

- Students will be aware of the importance of using scholarly sources and original material for verifying popular accounts of history.

Directions

Introduce the activity by summarizing the first two paragraphs of the overview. Distribute the readings and worksheet to students and have them work individually or in small groups. After they've had time to finish the worksheet, have the class discuss what conclusions they came to.

Readings

Charles Darwin. *On the Origin of Species.* 6th ed. (1872). Chapter 13.
  http://digital.library.upenn.edu/webbin/gutbook/lookup?num=2009

Charles Darwin. *The Voyage of the Beagle.* 2nd ed. (1845). Chapter 17.
  http://digital.library.upenn.edu/webbin/gutbook/lookup?num=3704

Adrian Desmond and James Moore. *Darwin: The Life of a Tormented Evolutionist.* New York: W.W. Norton & Company, 1991. Chapters 14-15.

Frank J. Sulloway. "Darwin's Conversion: The *Beagle* Voyage and Its Aftermath," *Journal of the History of Biology* 15 (1982), 1-53.

Frank J. Sulloway. "The Legend of Darwin's Finches." *Nature* 303 (1983), 372.

Frank J. Sulloway, "Darwin and the Galapagos," *Biological Journal of the Linnean Society* 21 (1984), 29-59.

Darwin and His Finches

It is widely believed that Darwin's theory of evolution was at least partly inspired by the finches he collected while he visited the Galápagos Islands, off the coast of Ecuador. This is the story told in the first episode of the *Evolution* video series. But is this what really happened?

Directions

Using the three books listed at the bottom of this worksheet, try to piece together a picture of what it was that helped inspire Darwin's theory of evolution. Then answer the two questions below.

1. What are some of the things that may have helped inspire Darwin's theory of evolution?

2. What are the reasons for your conclusions?

Readings

Charles Darwin. *On the Origin of Species*. 6th ed. (1872). Chapter 13.
   http://digital.library.upenn.edu/webbin/gutbook/lookup?num=2009
Charles Darwin. *The Voyage of the Beagle*. 2nd ed. (1845). Chapter 17.
   http://digital.library.upenn.edu/webbin/gutbook/lookup?num=3704
Adrian Desmond and James Moore. *Darwin: The Life of a Tormented Evolutionist*. New York: W.W. Norton & Company, 1991. Chapters 14-15.

3

Tracking Biology's Big Bang

Accompanies the *Viewer's Guide*, Chapter 2-D, "The Cambrian Explosion"

Overview

Episode Two of the *Evolution* series introduced students to the Cambrian Explosion, sometimes called the Biological Big Bang. In a relatively brief period of time many new animal forms appeared in the fossil record. But just how big of a bang was it? Was it a firecracker pop, a nuclear blast, or something in between? And what does it mean for contemporary evolutionary theory? That's what your students will investigate in this activity.

Learning Objectives

- Students will gain an appreciation of the great diversity that characterizes the animal kingdom.

- Students will be aware of when various phyla first appeared in the fossil evidence.

- Students will understand some of the implications of the Cambrian Explosion for contemporary evolutionary theory.

Directions

This activity is a class research project in which students will gather information about animal phyla and their first appearance in the fossil record. When they have gathered and compiled the information, they will be asked to compare their information about when the phyla first appeared and the predictions of contemporary Darwinian theory.

Introduce the activity by telling students they will be investigating the Cambrian Explosion, mentioned in Episode Two of the *Evolution* series. Hand out the instruction sheet, report form, filled-in sample report form, and list of references on the following pages.

Go over the instruction sheet with students, introducing the concept of the phylum and going over the directions for the activity and for filling out the report. Tell them that they should be prepared to orally present the results of their research to the rest of the class. If students want, they can include pictures with their presentation—which might help other students understand what representative members of the different phyla look like. Emphasize, however, that the presentations will be very short, and limited to the information contained in their report forms. Set a due date for the reports that seems realistic for your students and your schedule.

At the beginning of the class period when students will be presenting their reports, draw a table (chart) on the black board. Along the left side of the table, list the phyla in alphabetical order. Along the top of the table, from left to right, list the following geological periods: Precambrian, Cambrian, Ordovician, Silurian, Devonian, Carboniferous, Permian, Triassic, Jurassic, Cretaceous, Tertiary, Quaternary. The Precambrian is the earliest time period, while the Quaternary is the latest. (The Precambrian is actually an geological *era*. For the purposes of this activity, however, the distinction is unimportant.)

We have provided an example of such a table on the page with the title: "Tracking Biology's Big Bang: First Appearances (Sample)." Note that this is for illustrative purposes only and does not contain all the phyla or geological periods.

As students give their presentations, have them plot the first appearance of the phyla they report on. Let them know that the "lower" part of a period is the earliest, and the "upper" part is the most recent.

When the reports have been given and the results plotted, hand out the sheet titled, "What would Darwin predict?" and have the class discuss the questions it raises.

Tracking Biology's Big Bang

Episode Two of the *Evolution* series introduced you to the Cambrian Explosion, sometimes called the Biological Big Bang. In a relatively brief period of time many new animal forms appeared in the fossil record. But just how big of a bang was it? Was it a firecracker pop, a nuclear blast, or something in between? And what does it mean for Darwin's theory of evolution? That's what your class will investigate in this activity.

Each student in your class will be assigned an animal phylum to report on. A phylum (phyla for plural) is the broadest classification of animals. As opposed to a single species, like a chimpanzee, a horseshoe crab, or a horsefly, a phylum takes in a wide variety of animals. The phylum that contains humans also contains elephants, squirrels, canaries, lizards, guppies, and lampreys. Indeed, it contains every animal with a backbone—and then some.

If the differences within a phylum are great, the differences between phyla are vast. As much as a chimpanzee may differ from a fish, it differs even more radically from a sea urchin or a worm. In fact, you could say it's built on an entirely different architectural theme, or body plan.

Because phyla are so different from each other, the appearance of new phyla in the fossil record tells us something about how fast change is happening—depending, of course, on how good the fossil record is. The appearance of a new phylum every few million years or so might indicate gradual, steady change. The appearance of several new phyla at once, however, could indicate that something very different was going on.

Directions

Below is a list of animal phyla. You will be asked to report on one of them.

Acanthocephala	Annelida	Arthropoda
Brachiopoda	Bryozoa/Ectoprocta	Chaetognatha
Chordata	Cnidaria	Ctenophora
Echinodermata	Echiura	Entoprocta
Gastrotricha	Gnathostomulida	Hemichordata
Kinorhyncha	Loricifera	Mollusca
Nematoda	Nematomorpha	Nemertea
Onychophora	Phoronida	Placozoa

Platyhelminthes	Pogonophora	Porifera
Priapulida	Rotifera	Sipuncula
Tardigrada		

The information you'll need to get will be specified on a form that the teacher will hand out. The form will ask for a description of the phylum, the names of some animals that belong in the phylum, and the period that the phylum first appeared in the fossil record. (Note, though, that some phyla alive today do not appear in the fossil record. As part of your investigation, you'll be asked to indicate whether this is the case.) You may also be asked to include pictures of some of the animals that belong in the phylum.

Your teacher will give you a list of suggested sources where you can find the information you need. Once you and your classmates are done gathering information, you will plot the appearance of each phylum on a chart to see the pattern that emerges from the fossil evidence.

Tracking Biology's Big Bang

Phylum Report Form

Phylum Name:

Description:

Examples of animals belonging in this phylum: (Give no more than three.)

1.

2.

3.

Do any members of this phylum appear in the fossil record? (Check one.)

Yes __   No __

According to conventional dating, when did it first appear in the fossil record? (Use the geological period—for example, the lower Permian, upper Cambrian, etc.)

Where did you get your information about this phylum?

Tracking Biology's Big Bang (Sample)

Phylum Report Form

Phylum Name: Arthropoda

Description:

This phylum contains animals that have segmented bodies, skeletons on the outside of their bodies, and jointed appendages (i.e. legs, pincers, mouthparts, etc.)

Examples of animals belonging in this phylum: (Give no more than three.)

1. Crabs

2. Lobsters

3. Insects

Do any members of this phylum appear in the fossil record? (Check one.)

Yes <u>X</u> No __

According to conventional dating, when did it first appear in the fossil record? (Use the geological period—for example, the lower Permian, upper Cambrian, etc.)

The first appearance of the phylum was in the lower Cambrian.

Where did you get your information about this phylum?
Simon Conway Morris. "The Cambrian 'explosion': Slow-fuse or megatonnage?" *Proceedings of the National Academy of Sciences* 95, no. 9 (2000), 4426-4429. http://www.pnas.org/cgi/reprint/97/9/4426.pdf

Tracking Biology's Big Bang

Sources

You can find a great deal of information about animal phyla in paleontology and introductory biology textbooks. The best place to find these texts is in your local college or university library, but you can also find them in your local public library or school library. The more recent the texts, the better.

You can also find useful information on the Web—though the quality of information can vary. The best information is usually found on Web pages that document their facts with references to articles in professional journals. To search for information on the Web, go to Google (www.google.com) or any other large search engine. Use the name of the phylum you've been assigned as a search term. You can use it alone or with other search terms, such as "paleontology," "fossil," and so on.

Web

The following Web sites have information on a large number of animal phyla, as well as pictures of representative species:

Phyla of the Animalia (Lamont-Doherty Earth Observatory, Columbia Univ.)
> http://www.ldeo.columbia.edu/dees/ees/life/slides/oldec/animalia.00.html

General Overview of Animal Phyla (Bellarmine University)
> http://cas.bellarmine.edu/tietjen/images/general_overview_of_animal_phyla.htm

Print

The following printed sources may also be helpful:

Lynn Margulis and Karlene V. Schwartz, eds. *Five Kingdoms: An Illustrated Guide to the Phyla of Life on Earth.* 3rd ed. New York: W.H. Freeman and Co., 1998.

Simon Conway Morris and H. B. Whittington. "The Animals of the Burgess Shale." *Scientific American* 241 (1979), 122-133.

Simon Conway Morris. *The Crucible of Creation.* Oxford: Oxford University Press, 1998.

Simon Conway Morris. "The fossil record and early evolution of the Metazoa." *Nature* 361 (1993): 219-225.

Simon Conway Morris. "The Cambrian 'explosion': Slow-fuse or megatonnage?" *Proceedings of the National Academy of Sciences* 95, no. 9 (2000): 4426-4429.
> http://www.pnas.org/cgi/reprint/97/9/4426.pdf or
> http://www.pnas.org/cgi/content/full/97/9/4426

Jeffrey S. Levinton. "The Big Bang of Animal Evolution." *Scientific American* 267 (1992): 84-91.

Keep in mind that the classifications of some animals have changed over the years. For example, some classes of animals thought to belong to one phylum have been recognized as phyla in their own right.

Tracking Biology's Big Bang: First Appearances (Sample)

	Precambrian	Cambrian	Ordovician	Silurian	Devonian	Carboniferous	Permian	Triassic
Annelida								
Arthropoda								
Brachiopoda								
Bryozoa								
Chaetognatha								
Chordata								
Cnidaria								
Ctenophora								
Echinodermata								
Hemichordata								
Mollusca								
Nematoda								
Onychophora								
Phoronida								
Platyhelminthes								
Pogonophora								
Porifera								
Rotifera								

Tracking Biology's Big Bang

What Would Darwin Predict?

Contemporary Darwinism holds that all living organisms descended from a single "universal" ancestor. All the plants, animals, and other organisms that exist today are products of random changes and natural selection.

According to contemporary Darwinism, nature acts like a breeder, carefully scrutinizing every organism. As useful new traits occur, they are preserved and passed on to the next generation, while harmful traits are eliminated.

Though each change is small, these changes eventually accumulate to produce new tissues, organs, limbs or other parts. Given enough time, organisms may change so radically that they bear almost no resemblance to their original ancestor—or to their distant cousins alive today. Thus, humans, squids and dragonflies differ dramatically from their alleged single-celled ancestor. And they differ as dramatically from each other.

If we were to chart the appearance of new animal phyla, the chart would look something like the following diagram. As animals slowly diversify by accumulating changes, more and more new phyla begin to emerge, but it's a long, gradual process.

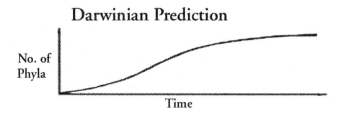

How do you think this compares with what your class found about the first appearance of the animal phyla? As you look at the chart you filled out in class, do you see increasing numbers of phyla appearing over time? Or do you see another pattern? If so, what kind of pattern do you see? Does it conflict with contemporary Darwinism? Why or why not?

A C T I V I T Y

4

The Centrality of Evolution

Accompanies the *Viewer's Guide,* Conclusion, Section D, "Is Evolution Indispensable to Medicine, Agriculture, and Choice of Mate?"

Overview

In a 1973 article in *The American Biology Teacher,* geneticist Theodosius Dobzhansky made the now-famous statement that "nothing in biology makes sense except in the light of evolution." This theme appears throughout the *Evolution* video series, which portrays Darwin's theory of evolution as central to both medicine and agriculture.

Yet even some scientists friendly to this message think Dobzhansky may have overstated his case. Molecular biologist Bruce R. Levin, for example, has noted, "While evolution may well be the thread that ties all of biology together, concern about the fabric of the subject seems to have had little play in much of modern biology. There are professional biologists who would be indifferent to the .. substance of Theodosius Dobzhansky's 1973 essay. . . . Indeed, as I found the other day, when speaking with a bright, and not-that-young, molecular geneticist, there are biologists out there who have never heard of Professor Dobzhansky. One can be a successful practitioner of many areas of contemporary biology without considering how organisms, molecules or phenomena came to be or their descent relationships. A relative absence of interest in evolution prevails in a number of areas of biology, with high-tech molecular biology being the most prominent among them."[1]

In a similar vein, Peter Grant, whose landmark work on the Galápagos finches was featured in Jonathan Weiner's Pulitzer Prize-winning book, *The Beak of the Finch,* stated in his 1999 presidential address to the American Society of Naturalists, "Not all biologists who would call themselves naturalists pay attention to [Dobzhansky's maxim] or even feel the need to. For example, an ecologist's world can make perfect sense, in the short term at least, in the absence of evolutionary considerations."[2]

Just how important is Darwin's theory of evolution to such areas as agriculture and medicine? One way to find out is to ask practitioners in each field what they think. In this activity, students will interview farmers or physicians to discover their views on how important Darwin's theory is to their work.

Note: In this activity students will be constructing and administering a questionnaire. Make sure students understand that they are not conducting a scientific survey. Although the survey may tell students what their respondents think about evolution's centrality, it won't necessarily tell them what doctors and farmers as a whole think.

Learning Objective

- Students will know how some farmers or doctors in their community view the importance of evolution to their work.

Directions (will require multiple class sessions)

Introduce this activity by referring to those segments of the series that talk about the importance of evolution for agriculture and medicine. (Those segments include Episode One on HIV; Episode Three on leafy spurge; Episode Four on tuberculosis and cholera; and Episode Six on diabetes—though you may only want to refer to those segments that they have already seen.) Tell the class that they are going to investigate what doctors or farmers think about the importance of evolution to their work. They are going to do this by creating a brief, informal questionnaire and then using it to interview the doctors or farmers.

Take the students through the following steps (spread these steps over as many class periods as you need):

1. Decide with the class whether they will interview doctors or farmers or both. (To keep the task of designing a questionnaire as simple as possible, it is recommended that you choose either doctors or farmers rather than both.) When you pick a group, make sure that there are enough so that each student can interview one person.

2. Brainstorm some questions for the questionnaire. A good questionnaire will take "multiple routes" to get to the same information. Try to come up with three to five different questions that get at the information you need. Below are some sample questions for doctors that ask about the importance of evolution in different ways:

   - In medical school, how many classes did you take on evolution?

   - In medical school, how much was evolution integrated into your classes?

   - How would your practice of medicine be different if you were unaware of Darwin's theory of evolution?

   - How much of a role does Darwin's theory of evolution play in your work?

   - On a scale of 1 to 5, with 1 being unimportant and 5 being extremely important, rate the importance of evolution to your practice of medicine.

Unimportant		Moderately Important		Extremely Important
1	2	3	4	5

3. Agree on single set of procedures for administering the questionnaire. Will you administer the questionnaire by phone or in person? How will the students introduce themselves and explain what they're doing? Should students formulate a script to read?

4. Assign a student to assemble the questions and procedural instructions into a single form.

5. Decide with the class how they will identify and select people to interview. Make sure that whatever procedure you choose guarantees that no respondent will be approached by more than one student. Since this isn't a scientific survey, don't worry about trying to obtain a random survey. Also, be sure to assign each student more than one person to interview, because not all respondents will be willing or able to be interviewed.

6. Assign one or more students to compile a list of respondents who will be interviewed by students.

7. When the questionnaire form and respondent list are completed, pass out the forms and assign each student some people to interview.

8. After the students have had time to complete their interviews, tally the responses to each question in class and discuss the class's findings. Possible questions to ask include:

   • What do the results of the interviews say about the importance of evolution for medicine/agriculture? Explain your answer.

   • For those who said that evolution is unimportant, is it likely that they were simply unaware of how evolution affects their work? Why or why not?

   • If we can look in such fields as medicine or agriculture and find examples of natural selection, does that mean evolution is vital to that field? Why or why not?

Notes

1. Bruce R. Levin, "Science as a Way of Knowing—Molecular Evolution." *American Zoologist* 24 (1984), 541-464.

2. Peter R. Grant, "What Does It Mean to Be a Naturalist at the End of the Twentieth Century?" *The American Naturalist* 155 (2000), 1-12.

5

How Do You Know?

Accompanies the *Viewer's Guide,* Chapter Four, "The Evolutionary Arms Race"

Overview

An important aspect of scientific literacy is being able to judge how much confidence we can place in the conclusions that we draw from empirical research or other sources. Part of this skill involves being able to spot the difference between well-supported claims and conjectural ones—those that are founded on insufficient information and reasoning.

One kind of speculation is what Harvard paleontologist Stephen Jay Gould, featured in Episode One, has called "just-so stories."

"Evolutionary biology," Gould said, "has been severely hampered by a speculative style of argument that records anatomy and ecology and then tries to construct historical or adaptive explanations for why this bone looked like that or why this creature lived here."

Gould continues, "Scientists know that these tales are stories; unfortunately, they are presented in the professional literature where they are taken too seriously and literally. Then they become 'facts' and enter the popular literature, often in such socially dubious form as the ancestral killer ape who absolves us from responsibility for our current nastiness, or as the 'innate' male dominance that justifies cultural sexism as the mark of nature."[1]

In other words, what sometimes happens is that biologists will gather information about an animal's current traits and environment—but then go beyond the data by inventing a scenario to explain how things got the way they are.

Although Episode Four does not present us with any ancestral killer apes, it does present us with conjectural scenarios. And it does so without warning viewers. In this activity, students will view Episode Four and try to spot such scenarios.

Learning Objectives

- Students will be aware that scientific statements may have varying levels of support.

- Students will be aware that even scientists may present conjectural claims as factual.

• Students will identify conjectural historical claims made in Episode Four.

Directions (will require multiple class sessions)

This activity will require you to show the first half of the episode during one class session and the second half during the next class session.

First Session

At the beginning of the first class, tell students that they will be viewing Episode Four of the *Evolution* series. However, instead of viewing the video the way they have in the past, students will be looking for certain kinds of things. In particular, they will be looking for examples of conjecture.

Ask students if they know what the term "conjecture" means. Have one or two volunteers define the term. Acceptable definitions should identify conjecture as a conclusion or opinion that is based upon insufficient evidence. Some students might have the idea that a conjecture is something that's untrue. Correct this misconception by pointing out that a conjecture may be true, but that it's unsupported by sufficient evidence.

Use the following questions to help flesh out the idea of what conjecture is:

Q. If we wanted to know how fast cheetahs can run, how would we go about finding that out?

A good answer would be something like, "We clock the speed of cheetahs when they are chasing game."

Q. Now, let's say someone measures the speeds of cheetahs in different parts of Africa. On the basis of his measurements, he then says that the cheetahs got their great speed due to environmental pressures that forced them to run fast to catch game and survive. Is this a well-supported claim or conjecture?

Students should identify this as conjecture. Measuring the running speed of a population of cheetahs at one point in time cannot tell us whether the speed of cheetahs has increased over several generations. That kind of information requires longitudinal research. We would have to measure the speed of cheetah populations over many years. We would also need observations to understand what environmental pressures—if any—were causing change. One way to get an estimate of cheetah speed over the years would be if paleontologists obtained fossils of cheetahs and inferred their speed from the fossilized cheetahs anatomical structure. But in that case, we'd also have to determine how well the fossils allow us to estimate the cheetah's speed.

Instruct students to jot down any claims in the video that they think are conjecture. Tell them that the next time the class meets they should be prepared to explain why they think those claims are conjectural.

Show the first half of the video.

Second Session

At the beginning of the second class session, show the second half of the video.

After the video, have students say which parts of the video they thought were conjectural and discuss why they think so. Remind students that conjecture isn't necessarily false, just unsupported. To keep them focused on this distinction, have students discuss how scientists could go about testing the conjectural statements.

Notes

1. Stephen Jay Gould, "Introduction," in Björn Kurtén, *Dance of the Tiger: A Novel of the Ice Age* (New York: Random House, 1980), xvii-xviii.

ACTIVITY

6

The Scopes Trial in Fact and Fiction

Accompanies the *Viewer's Guide,* Chapter 7-A, "The Creation-Evolution
Controversy and U.S. Science Education"

Overview

Many accounts of the Scopes trial are heavily colored by the trial's portrayal in the play *Inherit the Wind.* In the past, social studies teachers have sometimes been encouraged to use *Inherit the Wind* to teach students about the Scopes trial, but in recent years historians have seriously questioned the historical accuracy of the play. *Inherit the Wind* presents the Scopes trial as a stark showdown between defenders of free speech and religious fundamentalists who want to censor science teachers with whom they disagree. *Inherit the Wind* further suggests that Scopes' opponents were motivated primarily by a desire to defend a literal reading of the Bible. In reality, the events of the Scopes trial and the motivations behind it were much more complex and varied. In this assignment, students will examine differences between the real Scopes trial and the fictional portrayal of the trial in *Inherit the Wind.* In the process, they will have the opportunity to explore how their perceptions of historical reality are shaped by films and television.

Learning Objectives

- Students will understand key differences between the real Scopes trial and the portrayal of the trial in *Inherit the Wind.*

- Students will be aware of how their perceptions of historical reality may be shaped by films and television.

Directions (will require multiple class sessions)

Introduce the assignment by showing one of the film or television versions of *Inherit the Wind* in class or by assigning students to watch it outside of class. (*Inherit the Wind* is readily available on video.)

After your students have watched the video, distribute the handout on the following page and go over it with the students. Assign a due date that works with your schedule.

For background reading on the Scopes Trial (and its later fictionalization in *Inherit the Wind*), an excellent resource is the Pulitzer Prize-wining book *Summer for the Gods: The Scopes Trial and America's Continuing Debate over Science and Religion* by historian Edward Larson (Basic Books, 1997).

The Scopes Trial in Fact and Fiction

In 1925 a teacher named John Scopes was put on trial for allegedly violating a law that restricted the teaching of human evolution in Tennessee's public schools. *Inherit the Wind* presents a powerful fictionalized account of the Scopes trial, an account that has shaped many people's perceptions of what the Scopes trial was about. Schools sometimes use *Inherit the Wind* to teach students about the history of the Scopes trial. But what is fact and what is fiction in *Inherit the Wind*? And are there any drawbacks to relying on a fictionalized account learn about an historical event?

Read the sources listed below and write a brief report that answers the following questions:

1. According to *Inherit the Wind*, what was the Scopes trial about?
2. Describe the motivations of those opposed to the teaching of evolution in *Inherit the Wind*.
3. According to Carol Iannone, what are some key differences between the real Scopes trial and its fictionalized counterpart in *Inherit the Wind*?
4. Which of the differences noted by Iannone are significant in your view? Why?
5. Did the writers of *Inherit the Wind* intend their play to be viewed as an accurate depiction of the real Scopes trial? What was the message they thought they were communicating in their play?
6. What possible problems are there in relying on movies and plays to teach us about historical events?
7. Why do you think *Inherit the Wind* remains popular today?

Sources

1. Carol Iannone, "The Truth about *Inherit the Wind*."
   http://www.firstthings.com/ftissues/ft9702/articles/iannone.html
2. "Notes on *Inherit the Wind*," Famous Trials in American History website
   http://www.law.umkc.edu/faculty/projects/ftrials/scopes/SCO_INHE.HTM
3. "*Inherit the Wind:* The Playwrights' Note."
   http://xroads.virginia.edu/~UG97/inherit/l&lnote.html
4. Interview with historian Edward Larson:
   http://beatrice.com/interviews/larson/

7

Who Are Darwin's Critics Now?

Accompanies the *Viewer's Guide,* Chapter 7-A, "The Creation-Evolution Controversy and U.S. Science Education"

Overview

Activity 1 of this series asked students to investigate why many of Darwin's contemporaries objected to his theory of evolution. This activity is similar—only now the focus will be on our own time. As in Darwin's time, Darwinian theory is opposed not only by biblical literalists, but also by a wide variety of religious believers and scientists. In this research project, students will learn about some of Darwin's current religious and scientific critics, and why they object to his theory.

Learning Objectives

- Students will be aware that Darwin's theory has drawn opposition from a broad diversity of critics.

- Students will be able to identify some of Darwin's scientific and religious critics, as well as some of the critics' objections to Darwin's theory.

Directions

Hand out the worksheet on the following two pages and go over worksheet directions with students.

Who Are Darwin's Critics Now?

Directions

The following people have voiced criticisms of Darwin's theory of evolution. They represent a wide range of perspectives, and some, such as Stephen J. Gould and Niles Eldredge, still consider themselves defenders of Darwinism.

David Berlinski	Michael Behe	William Dembski	Michael Denton
Niles Eldredge	Brian Goodwin	Stephen Jay Gould	Mae-Wan Ho
Phillip Johnson	Søren Løvtrup	Henry Morris	Colin Patterson
David Raup	Peter Saunders	Jonathan Wells	

Choose five people from the list above and answer the following questions about each:

- Who is he or she?

- What is his or her occupation?

- What is he or she best known for?

- What are his or her objections to Darwin's theory?

How do your answers compare to how opponents of evolution have been portrayed throughout the *Evolution* video series?

Sources

You can find information about these people on the Web and in a number of printed books and articles. Here are a few places to look.

Michael Behe. *Darwin's Black Box: The Biochemical Challenge to Evolution.* New York: Free Press, 1996.

David Berlinski. "The Deniable Darwin." *Commentary* 101 (June 1996). http://www.arn.org/docs/berlinski/db_deniabledarwin0696.htm

William A. Dembski. *Intelligent Design: The Bridge between Science and Theology.* Downers Grove, Ill: InterVarsity Press, 1999.

William A. Dembski, Ed. *Mere Creation: Science, Faith & Intelligent Design.* Downers Grove, Ill: InterVarsity Press, 1998.

Michael Denton. *Evolution: A Theory in Crisis.* Bethesda, Maryland: Adler and Adler, 1986.

Niles Eldredge and Stephen Jay Gould. "Punctuated Equilibria: An Alternative to Phyletic Gradualism." In T. J. M. Schopf, ed., *Models in Paleobiology,* 82-115. San Francisco: Freeman Cooper, 1972.

Niles Eldredge. *Reinventing Darwin: The Great Debate at the High Table of Evolutionary Theory.* New York: John Wiley & Sons, 1995.

Niles Eldredge. *Time Frames: The Evolution of Punctuated Equilibria.* Princeton, New Jersey: Princeton University Press,1985. (Originally published as Niles Eldredge. *Time Frames: The Rethinking of Darwinian Evolution and the Theory of Punctuated Equilibria.* New York: Simon and Schuster, 1985.)

Brian Goodwin. *How the Leopard Changed Its Spots: The Evolution of Complexity.* Princeton, N.J.: Princeton University Press, 1994.

Stephen Jay Gould. "Is a New and General Theory of Evolution Emerging?" *Paleobiology* 6 (1980), 1, 119-120.

Stephen Jay Gould and Niles Eldredge. "Punctuated Equilibria: The Tempo and Mode of Evolution Reconsidered." *Paleobiology* 3 (1977): 115-151.

Mae-Wan Ho and Peter T. Saunders. "Beyond Neo-Darwinism—An Epigenetic Approach to Evolution." *Journal of Theoretical Biology* 78 (1979): 573-91.

Phillip E. Johnson. *Darwin on Trial.* Downers Grove, Ill: InterVarsity Press, 1991.

Phillip E. Johnson. *The Wedge of Truth.* Downers Grove, Ill.: InterVarsity Press, 2001.

Søren Løvtrup. *Darwinism: The Refutation of a Myth.* London: Croom Helm, 1987. (Also published by Viking Penguin.)

Henry M. Morris. *Creation Science.* El Cajon, Calif.: Master Books, 2001.

Paul A. Nelson. "A Colin Patterson Sampler." *Origins & Design* 17:1.
http://www.arn.org/docs/odesign/od171/sampler171.htm

Paul A. Nelson. "Colin Patterson Revisits His Famous Question about Evolution." *Origins & Design* 17:1.
http://www.arn.org/docs/odesign/od171/colpat171.htm

David M. Raup. *Extinction: Bad Genes or Bad Luck?* New York: W.W. Norton & Company, 1992.

Jonathan Wells. *Icons of Evolution: Science or Myth?* Washington, DC: Regnery Publishing, 2000.

8

Teaching the Controversy: What's Legal?

Accompanies the *Viewer's Guide,* Chapter 7-C, "Controversy in a Public
High School" and Chapter 7-D, "The Lafayette School Board"

Overview

All too often, controversial subjects are marked by an abundance of sound
bites and a relative dearth of substantive, carefully reasoned debate. That's understandable, considering the public's demand for cogent, quickly digestible information. But saying that this is understandable is not the same as saying that it's
good. Sound bites are a very poor basis for understanding important issues—or
making important decisions.

In this exercise, students will be asked to dig deeper on the constitutional
issues related to teaching origins in public-school classrooms.

Learning Objectives

- Students will be able to describe some key scientific criticisms of Darwinism.

- Students will be able to assess whether presenting criticisms of Darwinism in
  public school classrooms violates the First Amendment.

- Students will be able to describe some core concepts of Intelligent Design
  theory and assess whether they are scientific or religious.

- Students will be able to assess whether presenting Intelligent Design in public
  school classrooms violates the First Amendment.

Directions

Go over the following handout with the students. Give them a due date that
works best with your schedule, taking into account your students' abilities. If you
wish, when the students have completed their reports, set aside a class session for
a class discussion/debate on the following questions:

- From what you've learned in doing your reports, were Mr. Spokes's plans to
  teach the origins controversy legally sound?

- Why or why not?

Teaching the Controversy: What's Legal?

All too often, controversial subjects are marked by too many sound bites and too few thoughtful arguments. That's not surprising, given the public's demand for dramatic, quickly digestible information. But is that a good way to understand important issues? Are sound bites a good basis for making important decisions?

In this assignment, you will get the chance to dig deeper on the question of how the origins controversy should be taught in the public schools. In an article titled, "Teaching the Origins Controversy" (see list of sources), the article's authors describe a fictitious science teacher who wants to make some changes in the way he teachers evolution in his classroom. The science teacher, John Spokes, wants to:

1) Correct blatant factual errors in his textbook that overstate the evidential case for neo-Darwinism.

2) Tell students about the evidential challenges to these theories that current textbooks fail to mention.

3) Define the term "evolution" without equivocation and to distinguish clearly between those senses of the term that enjoy widespread support among scientists and those that remain controversial, even if only among a minority of scientists.

4) Tell his students that a growing minority of scientists do see evidence of real, not just apparent, design in biological systems.

Using the article and the other sources listed at the end of this handout, write a report that answers the following questions:

Is It Science? Are Spokes's intended changes in his biology curriculum scientific? Is his plan to correct and critique textbook presentations of neo-Darwinism scientific? Are the alternative theories that Spokes wants to present (including the theory of intelligent design) scientific?

Is It Religion? Does Spokes's plan to correct and critique textbook presentations of neo-Darwinism constitute an establishment of religion? Does Spokes's plan to expose his students to evidence of design and design theory qualify as teaching religion? Does the First Amendment prevent the presentation of this point of view?

Is It Speech? Are Spokes's plans to correct and critique textbook presentations of neo-Darwinism, and to expose students to the alternative theory of intelligent design, protected under the First Amendment?

Sources

Use the following sources to answer the questions for this report. If you wish, you may also read through some of the United States Supreme Court decisions mentioned in these sources. A good place to find the text of these decisions is at www.findlaw.org. When you go to the site, enter the name of the case in the top left-hand text box, using all uppercase letters, except for the "v". For example, if you want to read the Supreme Court's decision on Louisiana's "balanced-treatment" law, you would enter "EDWARDS v. AGUILLARD" in the search box. Then select "US Supreme Court" from the pull-down menu box on the right and click the "Search" button.

David K. DeWolf, Stephen C. Meyer and Mark Edward DeForrest. "Teaching the Origins Controversy: Science, Or Religion, Or Speech?" *Utah Law Review* 39 (2000).

http://www.arn.org/docs/dewolf/utah.pdf

David K. DeWolf, Stephen C. Meyer, Mark E. DeForrest. *Intelligent Design in Public School Science Curricula: A Legal Guidebook.* Richardson, Texas: Foundation for Thought and Ethics, 1999.

http://www.arn.org/docs/dewolf/guidebook.htm

Creationism v. Evolution: Will Religion or Science Prevail? On *Justice Talking,* from National Public Radio. Audio debate between Prof. David K. DeWolf, of the Gonzaga School of Law in Spokane, Wash. and Eugenie C. Scott, of the National Center for Science Education in Oakland, Calif.

http://justicetalking.org/shows/show135.asp

CPSIA information can be obtained at www.ICGtesting.com
Printed in the USA
BVOW08s1148301215

431394BV00001B/22/P